NATURAL POLYMERS FOR PHARMACEUTICAL APPLICATIONS

Volume 1: Plant-Derived Polymers

Natural Polymers for Pharmaceutical Applications, 3-volume set:

Natural Polymers for Pharmaceutical Applications

Volume 1: Plant-Derived Polymers

Natural Polymers for Pharmaceutical Applications

Volume 2: Marine- and Microbiologically Derived Polymers

Natural Polymers for Pharmaceutical Applications

Volume 3: Animal-Derived Polymers

NATURAL POLYMERS FOR PHARMACEUTICAL APPLICATIONS

Volume 1: Plant-Derived Polymers

Edited by

Amit Kumar Nayak, PhD
Md Saquib Hasnain, PhD
Dilipkumar Pal, PhD

AAP APPLE
ACADEMIC
PRESS

Apple Academic Press Inc.   Apple Academic Press Inc.
3333 Mistwell Crescent   1265 Goldenrod Circle NE
Oakville, ON L6L 0A2   Palm Bay, Florida 32905
Canada   USA

© 2020 by Apple Academic Press, Inc.

First issued in paperback 2021

Exclusive worldwide distribution by CRC Press, a member of Taylor & Francis Group
No claim to original U.S. Government works

Natural Polymers for Pharmaceutical Applications, Volume 1: Plant-Derived Polymers
ISBN 13: 978-1-77463-182-9 (pbk)
ISBN 13: 978-1-77188-845-5 (hbk)
Natural Polymers for Pharmaceutical Applications, 3-volume set
ISBN 13: 978-1-77188-844-8 (hbk)
ISBN 13: 978-0-42932-812-1 (pbk)

Library and Archives Canada Cataloguing in Publication

Title: Natural polymers for pharmaceutical applications / edited by Amit Kumar Nayak,
 Md. Saquib Hasnain, Dilipkumar Pal.

Names: Nayak, Amit Kumar, 1979- editor. | Hasnain, Md Saquib, 1984- editor. |
 Pal, Dilipkumar, 1971- editor.

Description: Includes bibliographical references and indexes

Identifiers: Canadiana (print) 20190146095 | Canadiana (ebook) 20190146117 | ISBN 9781771888448
 (set ; hardcover) | ISBN 9781771888455 (v. 1 ; hardcover) | ISBN 9781771888462 (v. 2 ; hardcover) |
 ISBN 9781771888479 (v. 3 ; hardcover) | ISBN 9780429328121 (set ; ebook) | ISBN 9780429328251
 (v. 1 ; ebook) | ISBN 9780429328299 (v. 2 ; ebook) | ISBN 9780429328350 (v. 3 ; ebook)

Subjects: LCSH: Polymers in medicine. | LCSH: Biopolymers. | LCSH: Pharmaceutical technology.

Classification: LCC R857.P6 N38 2020 | DDC 615.1/9—dc23

CIP data on file with US Library of Congress

Apple Academic Press also publishes its books in a variety of electronic formats. Some content that appears in print may not be available in electronic format. For information about Apple Academic Press products, visit our website at **www.appleacademicpress.com** and the CRC Press website at **www.crcpress.com**

About the Editors

Amit Kumar Nayak, PhD

Amit Kumar Nayak, PhD, is currently working as an Associate Professor at the Seemanta Institute of Pharmaceutical Sciences, Odisha, India. He earned his PhD in Pharmaceutical Sciences from IFTM University, Moradabad, U.P, India. He has over 10 years of research experience in the field of pharmaceutics, especially in the development and characterization of polymeric composites, hydrogels, and novel and nanostructured drug delivery systems. To date, he has authored over 120 publications in various high-impact peer-reviewed journals and 34 book chapters to his credit. Overall, he has earned highly impressive publishing and cited record in Google Scholar (H-Index: 32, i10-Index: 80). He has been the permanent reviewer of many international journals. He also has participated and presented his research work at several conferences in India and is a life member of the Association of Pharmaceutical Teachers of India (APTI).

Md Saquib Hasnain, PhD

Dr. Md Saquib Hasnain has over 6 years of research experience in the field of drug delivery and pharmaceutical formulation analyses, especially systematic development and characterization of diverse nanostructured drug delivery systems, controlled release drug delivery systems, bioenhanced drug delivery systems, nanomaterials and nanocomposites employing Quality by Design approaches as well as development and characterization of polymeric composites, formulation characterization and many more. Till date he has authored over 30 publications in various high impact peer-reviewed journals, 35 book chapters, 3 books, and 1 Indian patent application to his credit. He is also serving as the reviewer of several prestigious journals. Overall, he has earned highly impressive publishing and cited record in Google Scholar (H-Index: 14). He has also participated and presented his research work at over 10 conferences in India, and abroad. He is also the member of scientific societies, i.e., Royal Society of Chemistry, Great Britain, International Association of Environmental and Analytical Chemistry, Switzerland and Swiss Chemical Society, Switzerland.

Dilipkumar Pal, PhD

Dilipkumar Pal, PhD, MPharm, Chartered Chemist, Post Doct (Australia) is an Associate Professor in the Department of Pharmaceutical Sciences, Guru Ghasidash Vishwavidyalaya (A Central University), Bilaspur, C.G., India. He received his master and PhD degree from Jadavpur University, Kolkata and performed postdoctoral research as "Endeavor Post-Doctoral Research Fellow" in University of Sydney, Australia. His areas of research interest include "Isolation, structure Elucidation and pharmacological evaluation of indigenous plants" and "natural biopolymers." He has published 164 full research papers in peer-reviewed reputednational and international scientific journals, having good impact factor and contributed 113 abstracts in different national and international conferences. He has written 1 book and 26 book chapters published by reputed international publishers. His research publications have acquired a highly remarkable cited record in Scopus and Google Scholar (H-Index: 35; i-10-index 82, total citations 3664 till date). Dr. Pal, in his 20 years research-oriented teaching profession, received 13 prestigious national and international professional awards also. He has guided 7 PhD and 39 master students for their dissertation thesis. He is the reviewer and Editorial Board member of 27 and 29 scientific journals, respectively. Dr. Pal has been working as the Editor-in-Chief of one good research journal also. He is the member and life member of 15 professional organizations.

Contents

Contributors

Mohammad Tahir Ansari
Department of Pharmaceutical Technology, Faculty of Pharmacy and Health Sciences,
Universiti Kuala Lumpur Royal College of Medicine Perak, Malaysia

Shanta Biswas
Department of Applied Chemistry and Chemical Engineering,
Faculty of Engineering and Technology, University of Dhaka, Dhaka – 1000, Bangladesh

Phool Chandra
School of Pharmaceutical Sciences, IFTM University, Lodhipur Rajput,
Delhi Road (NH-24), Moradabad–244102, UP, India

Devasish Chowdhury
Department of Material nanochemistry, Physical Science Division,
Institute of Advanced Study in Science and Technology (IASST),
Guwahati – 781035, Assam, India

Madhusmita Dhupal
Department of Global Medical Science, Wonju College of Medicine, Yonsei University,
Gangwon 220-701, Republic of Korea

Gaytri Gour
Pharmaceutics Research Projects Laboratory, Department of Pharmaceutical Sciences,
Dr. Hari Singh Gour Central University, Sagar (M.P.) – 470 003, India

Jian Guan
School of Pharmacy, Shenyang Pharmaceutical University, 103, Wenhua Road,
Shenyang, 110016, China

Mukesh Kumar Gupta
Department of Biotechnology and Medical Engineering, National Institute of Technology,
Rourkela, Odisha – 769 008, India

Papia Haque
Department of Applied Chemistry and Chemical Engineering,
Faculty of Engineering and Technology, University of Dhaka,
Dhaka – 1000, Bangladesh

Md Saquib Hasnain
Department of Pharmacy, Shri Venkateshwara University, NH-24, Rajabpur, Gajraula,
Amroha – 244236, U.P., India

Md Minhajul Islam
Department of Applied Chemistry and Chemical Engineering,
Faculty of Engineering and Technology, University of Dhaka,
Dhaka – 1000, Bangladesh

Ankit Jain
Institute of Pharmaceutical Research, GLA University, NH-2, Mathura-Delhi Road,
Mathura (U.P.) – 281406, India

Sanjay K. Jain
Professor of Pharmaceutics, Head, Department of Pharmaceutical Sciences,
Dean, School of Engineering and Technology, Pharmaceutics Research Projects Laboratory,
Department of Pharmaceutical Sciences, Dr. Hari Singh Gour Central University,
Sagar (M.P.) – 470 003, India

Mohit Kumar
School of Advanced Materials Science and Engineering, Sungkyunkwan University,
Suwon, South Korea

Abul K. Mallik
Department of Applied Chemistry and Chemical Engineering,
Faculty of Engineering and Technology, University of Dhaka,
Dhaka – 1000, Bangladesh

Shirui Mao
School of Pharmacy, Shenyang Pharmaceutical University, 103, Wenhua Road,
Shenyang, 110016, China

Sitansu Sekhar Nanda
Department of Chemistry, Myongji University, Yongin, South Korea

Amit Kumar Nayak
Department of Pharmaceutics, Seemanta Institute of Pharmaceutical Sciences,
Mayurbhanj – 757086, Odisha, India

Dilipkumar Pal
Department of Pharmaceutical Sciences, Guru Ghasidas Vishwavidyalaya (A Central University),
Koni, Bilaspur – 495009, C.G., India

Pritish Kumar Panda
Pharmaceutics Research Projects Laboratory, Department of Pharmaceutical Sciences,
Dr. Hari Singh Gour Central University, Sagar (M.P.) – 470003, India

M. Prabaharan
Department of Chemistry, Hindustan Institute of Technology and Science, Padur,
Chennai – 603103, India

Mohammed Mizanur Rahman
Department of Applied Chemistry and Chemical Engineering,
Faculty of Engineering and Technology, University of Dhaka,
Dhaka – 1000, Bangladesh

Neetu Sachan
School of Pharmaceutical Sciences, IFTM University, Lodhipur Rajput, Delhi Road (NH-24),
Moradabad–244102, UP, India

Md. Nurus Sakib
Department of Applied Chemistry and Chemical Engineering,
Faculty of Engineering and Technology, University of Dhaka,
Dhaka – 1000, Bangladesh

Shivani Saraf
Pharmaceutics Research Projects Laboratory, Department of Pharmaceutical Sciences,
Dr. Hari Singh Gour Central University, Sagar (M.P.) – 470003, India

D. Sathya Seeli
Department of Chemistry, Hindustan Institute of Technology and Science,
Padur, Chennai – 603103, India

Md Shahruzzaman
Department of Applied Chemistry and Chemical Engineering,
Faculty of Engineering and Technology, University of Dhaka,
Dhaka – 1000, Bangladesh

Sadia Sharmeen
Department of Applied Chemistry and Chemical Engineering,
Faculty of Engineering and Technology, University of Dhaka,
Dhaka – 1000, Bangladesh

Ankita Tiwari
Pharmaceutics Research Projects Laboratory, Department of Pharmaceutical Sciences,
Dr. Hari Singh Gour Central University, Sagar (M.P.) – 470003, India

Dipti Ranjan Tripathy
Department of Neurology, Gauhati Medical College and Hospital, Gauhati – 781032,
Assam, India

Amit Verma
Pharmaceutics Research Projects Laboratory, Department of Pharmaceutical Sciences,
Dr. Hari Singh Gour Central University, Sagar (M.P.) – 470003, India

Dong Kee Yi
Department of Chemistry, Myongji University, Yongin, South Korea

Abbreviations

AAc	acrylic acid
ALMP	amidated low-methoxyl pectin
Am.P	amidated pectin
AmB	amphotericin B
ASGPR	asialoglycoprotein receptor
ATRP	atom transfer radical polymerization
BMEP	bis [2-methacryloyloxy] and ethyl phosphate
BSA	bovine serum albumin
Ca.P	calcium salt of pectin
CAS/GT	casein/gum tragacanth
CC NPs	calcium carbonate nanoparticles
CRF	chronic renal failure
CS	chitosan
CZ	caseinate/zein
DE	degree of esterification
DET-CS NPs	de-esterified tragacanth-chitosan nanoparticles
DHA	dihydroartemisinin
DIP	dipyridamole
DNH	double network hydrogel
DS	diclofenac sodium
DSC	differential scanning calorimetry
EA	emulsifying activity
ECM	extracellular matrix
ES	emulsifying stability
FC	fermentable carbohydrate
FDA	Food and Drug Administration
FG	fenugreek galactomannan
FSG	fenugreek seed gum
GA	glutaraldehyde
GA	gum acacia
GA	gum Arabic
GAGP	arabinogalactan–protein complex
GA-MNP	gallic acid coated magnetic nanoparticles

GDE	glyceroldiglycidylether
GI	gastrointestinal
GIT	gastro-intestinal tract
GO	graphene oxide
GRAS	generally recognized as safe
GT	gum tragacanth
GT/PLLA	gum tragacanth-based poly (l-lactic acid)
GT-PVA	gum tragacanth-polyvinyl alcohol
HA	hydroxyapatite
HCPT	10-hydroxycamptothecin
HDL	high-density lipoprotein
HepG2	hepatocarcinoma cells
HG	homogalacturonan
HM	high-methoxyl
HMP	high methoxyl pectin
HNC	hydrogel nanocomposite
HPLC	high-performance liquid chromatography
HPMC	hydroxypropyl methylcellulose
IASST	Institute of Advanced Study in Science and Technology
IND	indomethacin
IP	isoelectric point
IPN	interpenetrating polymer networking
ITZ	itraconazole
KL	ketoprofen lysinate
LBG	locust bean gum
LDL	low-density lipoprotein
LM	low-methoxyl
LMP	low methoxyl pectin
MCDDS	microflora triggered colon targeted drug delivery system
MG53	mitsugumin 53
MWCNT	multi-walled carbon nanotube
OG	okra *(Hibiscus esculentus)* gum
PCL	poly (epsilon-caprolactone)
PDC	pectin-dihydroartemisinin
PEG	polyethylene glycol
PLGA	polylactic glycolic acid
PLLA/GT	poly (l-lactic acid) and gum tragacanth
PMVE/MA	poly (methyl vinyl ether-comaleic anhydride)

PT–ME	pectin–metronidazole
PTV	Punta Toro virus
PVA	poly (vinyl alcohol)
PVP	polyvinylpyrrolidone
PVP-I	polyvinylpyrrolidone iodine
QC	quercetin
RES	resveratrol
ROP	ring opening polymerization
SA	sodium alginate
SCF	simulated colonic fluid
SDS-PAGE	sodium dodecyl sulfate-polyacrylamide gel electrophoresis
SEM	scanning electron microscopy
SGF	simulated gastric fluid
SIF	simulated intestinal fluid
SM	silymarin
SPI	soy protein isolate
SWPI	soy whey protein isolate
TCH	tetracycline hydrochloride
TG	tragacanth gum
TmG	tamarind gum
TPGME	tripropylene glycol methyl ether
Trag	tragacanth
UA	ursolic acid
VI	N-vinyl imidazole
VLDL	very low-density lipoprotein
WPI	whey proteins
XRD	x-ray diffraction study

Preface

The sources of plant-derived polymers are abundant in the nature, and these natural polymers comprise a group of macromolecular polysaccharide biomaterials with a wider array of physicochemical behaviors. The utilization of the plant-derived polymers as potential pharmaceutical excipients is expanding day by day in terms of stability in the biological system, drug-releasing capability, drug targeting specificities at desired sites, and also the bioavailability of drugs. During the past few decades, many plant-derived polymers have been studied for the use as pharmaceutical excipients such as suspending agents, emulsifiers, binders, disintegrants, gelling agents, biomucoadhesive agents, matrix formers, release retardants, enteric resistants, etc., in various pharmaceutical dosage forms like tablets, microparticles, nanoparticles, ophthalmic preparations, gels, emulsions, suspensions, etc., and their applicability has been established for further industrial uses. The commonly used plant-derived polymers used as pharmaceutical excipients in different pharmaceutical dosage formulations are gum Arabica (GA), gum tragacanth (GT), pectin, guar gum, tamarind gum (TmG), sterculia gum, okra gum, etc.

This current volume of the book, *Natural Polymers for Pharmaceutical Applications, Volume I: Plant-Derived Polymers,* contains nine important chapters, which present the latest research updates on the plant-derived polymers for various pharmaceutical applications. The topics of the chapters of the current volume include but not limited to: pharmaceutical applications of TmG; pharmaceutical applications of GA; recent advances in pharmaceutical applications of natural carbohydrate polymer GT; application potential of pectin in drug delivery; guar gum and its derivatives: pharmaceutical applications; pharmaceutical applications of locust bean gum (LBG); pharmaceutical applications of sterculia gum; pharmaceutical applications of okra gum; and pharmaceutical applications of fenugreek seed gum (FSG). This book mainly discusses the aforementioned topics along with an emphasis on the recent advances in the fields by experts across the world.

We would like to thank all the authors of the chapters for providing timely and excellent contributions, the publisher Apple Academic Press

(USA), and Sandra Sickels for the invaluable help in the organization of the editing process. We gratefully acknowledge the permissions to reproduce copyright materials from a number of sources. Finally, we would like to thank our family members, all respected teachers, friends, colleagues, and dear students for their continuous encouragements, inspirations, and moral support during the preparation of the book. Together with our contributing authors and the publishers, we will be extremely pleased if our efforts fulfill the needs of academicians, researchers, students, polymer engineers, and pharmaceutical formulators.

—Amit Kumar Nayak, PhD
Md Saquib Hasnain, PhD
Dilipkumar Pal, PhD

CHAPTER 1

Pharmaceutical Applications of Tamarind Gum

AMIT KUMAR NAYAK,[1] SITANSU SEKHAR NANDA,[2] DONG KEE YI,[2] MD SAQUIB HASNAIN,[3] and DILIPKUMAR PAL[4]

[1]Department of Pharmaceutics, Seemanta Institute of Pharmaceutical Sciences, Mayurbhanj – 757086, Odisha, India

[2]Department of Chemistry, Myongji University, Yongin, South Korea

[3]Department of Pharmacy, Shri Venkateshwara University, NH-24, Rajabpur, Gajraula, Amroha – 244236, U.P., India

[4]Department of Pharmaceutical Sciences, Guru Ghasidas Vishwavidyalaya, Koni, Bilaspur – 495009, C.G., India

ABSTRACT

Presently, diverse plant polysaccharides contemplated for various applications as excipients like suspending agents, granulating agents, binders, disintegrating agents, mucoadhesives, gel-formers, release modifiers, enteric resistant, matrix formers, emulsifiers, etc., in several pharmaceutical dosage applications. From this, tamarind gum (TmG) is a coming forth excipient, which is being applied and looked into for the formulation of several dosage forms like emulsions, suspensions, tablets, creams, gels, beads, microparticles, spheroids, nanoparticles, buccal patches, and ophthalmic preparations, etc. The chapter messes with a comprehensive and utilitarian discourse on pharmaceutical diligences of TmG with its some significant features like isolation, source, and chemical composition and attributes (both physicochemical and biological).

1.1   INTRODUCTION

Polysaccharides are high molecular weight polymers possess branched and complex molecular structures with various monosaccharide residues, linked with the *O*-glycosidic linkages (Pal and Nayak, 2015, 2017). These are hydrophilic and gel-former biopolymeric materials (Nayak et al., 2010, 2012). The current socio-economic position of the modernistic world has raised the interestingness in the use of natural polysaccharides, as the replacement of different synthetic biopolymers in useful biomedical applications (Hasnain et al., 2010; Nayak and Pal, 2012, 2015, 2016a). Natural polysaccharides prevail from the plant, algal, microbial, and animal origins (Pal and Nayak, 2016). Numerous natural polysaccharides are produced *in vitro* by a biotechnological enzymatic process. Plant polysaccharides are a popular natural biopolysaccharide group, which are less expensive, non-toxic, biodegradable, and widely available in the natural sources (Hasnain et al., 2017a, b, 2018a, b; Nayak and Pal, 2016b, 2017). These occur mainly in fruits, seeds, exudates materials, roots, leaves, pods, rhizomes, etc. (Nayak et al., 2018a, b, c; Prajapati et al., 2013). The fact of the increasing significance of the use of plant polysaccharides is related with its harvesting or cultivating in a sustained manner and able to offer a constant supply of raw materials (Nayak and Pal, 2013). The use of plant polysaccharides in pharmaceutical applications including drug delivery is developing from their traditional auxiliary function in formulations, and its dynamic roles as drug delivery excipient agents deal with stability, drug release, target specific therapeutic action, bioavailability (Nayak and Pal, 2017). Currently, an enormous number of plant polysaccharides have been produced from a number of ordinarily available local plants (Nayak et al., 2013a, b, c, 2014a, b, 2015; Sinha et al., 2015a). Even these plant polysaccharides have been employed in the formulation of numerous kinds of pharmaceutical products as excipients (Bera et al., 2015a, b, c; Das et al., 2013; Guru et al., 2013, 2018; Jena et al., 2018; Malakar et al., 2013; Prajapati et al., 2013; Sinha et al., 2015b). Among various plant polysaccharides, TmG is the galactoxylan, which is extracted from the tamarind kernel and have found its wide therapeutic actions in the pharmaceutical fields such as foods and cosmetics (Nayak, 2016; Shaikh et al., 2015). Recent years, TmG is practiced as necessitate pharmaceutical ingredients in diverse categories of dosage applications. This chapter relates with a useful and complete discourse on the pharmaceutical applications of TmG. Besides this, some significant features of TmG like isolation, source, chemical composition, and attributes (both physicochemical and biological) are also discussed in brief.

1.2 SOURCES AND ISOLATION

Tamarind (*Tamarindus indica*, family: Fabaceae) tree, commonly recognized as 'Indian date,' is an evergreen tree. Tamarind is cultivated throughout the whole of India and also, in other Southeast Asian countries (Nayak, 2016). Tamarind seed incorporates the endosperm or the kernel (70–75%), seed testa or coat (20–30%). Tamarind seeds contain 67.1 g/Kg crude fiber and huge contents of carbohydrates (Joseph et al., 2012). TmG is a cell wall storage material, present in the seed, and is separated from its seed powder (Nayak et al., 2016). The isolation method of TmG was first derived in the laboratory by Rao and Ghosh (1946). The procedure was improved by Rao and Srivastava (1973) and further modified by Nandi (1975) on the laboratory scale. Several researchers have demonstrated some other procedures of TmG isolation. Mainly, these procedures can be classified as chemical methods and enzymatic methods. In the chemical method, tamarind kernel powder is boiled with water. Then, the extracted mucilage is added to an equal amount of acetone to produce precipitation, which is collected and then, dried (Nayak et al., 2015). The dried powder is mixed with ethanol, processed with the enzyme protease and subsequently, centrifuged. The remaining supernatant portion is added to ethanol. The solution is dried and separated (Tattiyakul et al., 2010).

1.3 CHEMICAL COMPOSITION AND PROPERTIES

Chemically, TmG is a highly branched polysaccharide framed of $(1 \rightarrow 4)$-β-D-glucan backbone exchanged with the side chains of α-D-xylopyranose and β-D-galactopyranosyl $(1 \rightarrow 2)$-α-D-xylopyranose connected $(1 \rightarrow 6)$ to glucose remainders (Figure 1.1) (Nayak and Pal, 2011). Xylose remainders (1–6 linked) were replaced by almost 80% of glucose and partially substituted by the p-1-2 galactose residue (Manchanda et al., 2014). The molecular structure of TmG comprises some monomers of glucose, galactose, and xylose monomer units in 2.80: 1.00: 2.25 molar ratios (Kaur et al., 2012a). Thus, TmG is regarded as a galactoxyloglucan (Jana et al., 2013). It is biocompatible, noncarcinogenic, and stable in the acidic pH (Manchanda et al., 2014; Nayak, 2016). However, it is insoluble in the organic solvents and in the cold water (Nayak et al., 2015, 2016). Native TmG exhibits a tendency of aggregation when aqueous solvents take part in circulation. The aggregates dwell of lateral fabrications of single polysaccharide filaments showing behavior. This can be comfortably described by Kuhn's model or by the worm-like chain (Joseph et al., 2012). It has an excellent ability to swell in the aqueous medium and produces the mucilaginous solution (Kaur et al., 2012). Therefore, it has hydrophilic, gel-forming, and bioadhesive properties (Pal and Nayak, 2012). Because of these properties, it is also in preparation of hydrogels (Meenakshi and Ahuja, 2015). TmG is found non-irritant with hemostatic activity (Avachat et al., 2011). It has shown hepatoprotective activity, also (Samal and Dang, 2014).

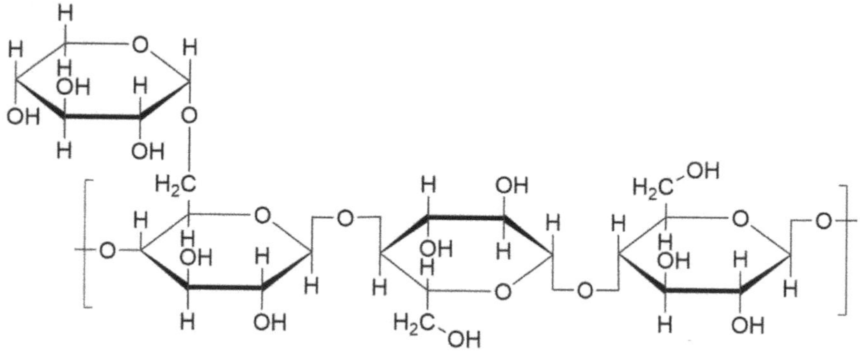

FIGURE 1.1   Chemical structure of TmG.

1.4   PHARMACEUTICAL APPLICATIONS

Presently, a diverse kinds of plant polysaccharides contemplated for various applications as pharmaceutical and drug delivery excipients like suspending agents, granulating agents, binders, disintegrating agents, mucoadhesives, gel-formers, release modifiers, enteric resistant, matrix formers, emulsifiers, etc., in several dosage forms (Prajapati et al., 2013; Biswas and Sahoo, 2016; Nayak and Pal, 2017; Pal et al., 2010). Among these plant polysaccharides, TmG is emerging as a potential biopolymeric excipient material for the pharmaceutical applications. The use of TmG for the preparation of diverse pharmaceutical dosage forms is discussed below.

1.5 EMULSIONS AND SUSPENSIONS

TmG was also studied as an emulsifier in various preparations of emulsions. In an investigation by Kumar et al., (2001), a comparative study on the castor oil emulsions with using TmG and gum acacia (GA) (a commonly used emulsifier as standard) as emulsifiers demonstrated the effectiveness of 2% w/v of TmG than 10% w/v of GA.

TmG was also exploited as suspending agents in various pharmaceutical suspensions (Deveswaran et al., 2010; Malaviya et al., 2010). In these fields, researchers have found the suitability of TmG as suspending agent to produce stable suspensions. It was found to reduce the settling rate of solid particles of these prepared suspensions and to permit also in the easy redispersion of any settled particles (Deveswaran et al., 2010). In an investigation, the suspending characteristics of TmG in a paracetamol suspension were compared with some ordinarily applied suspending agents like tragacanth (Trag), gelatin, and Arabica gum (Malaviya et al., 2010). From these investigations, a promise in the use of TmG as a suspending agent in the pharmaceutical suspension was indicated.

1.6 TABLETS

TmG was already studied as excipients like binders, matrix formers, and release modifiers in various kinds of pharmaceutical tablet formulations. It was investigated as an effective binder material for the weight granulation and the direct compression in various tablets (Kulkarni et al., 2011). When the tablet binding character of TmGin pharmaceutical tablets for various types of drugs was studied, it was observed that these tablets exhibited slower drug release profiles. This was attributed to the hydrophilicity, viscosity, and higher swelling of TmG.

TmG was also studied as matrix formers in the matrix tablets containing drugs of various molecular weights (Joseph et al., 2012). In most of the cases, matrix tablets are formulated to make sustained drug releasing or controlled drug releasing formulations for which these require release modifiers or release retardants. Owing to its hydrophilic characteristics, TmG was extensively used in various matrix tablets as matrix formers as well as release retardants (Chanda and Roy, 2010; Km and Arul, 2016; Mali and Dhawale, 2016; Phani Kumar et al., 2011). Along with the sustained drug-releasing pattern, some matrix tablets composed of TmG exhibited a mucoadhesive property, which was found helpful in the gastroretentive drug delivery (Mali and Dhawale, 2016; Rajab et al., 2012). Table 1.1 presents some examples of tablets in which, TmG was used as binders, matrix formers, release retardants, and mucoadhesive agents.

Recently, some investigations were performed by various researchers, where modified forms of TmG were exploited with the matrix materials in the preparation of sustained drug releasing matrix tablets for numerous drugs (Ghosh and Pal, 2013; Jana et al., 2014; Sravani et al., 2011). Ghosh and Pal (2013) investigated the formulation of aspirin matrix tablets using polyacrylamide-grafted TmG. These matrix tablets demonstrated the controlled drug releasing (zero-order) behavior. It has also been found that the rate of drug releasing pattern was decreasing with an increment of % grafting. The drug release was found lower in the acidic pH and was much more prominent in the neutral pH as well as in the alkaline pH. Sravani et al., (2011) formulated diclofenac sodium (DS) and ketoprofen matrix tablets using TmG and epichlorohydrin-cross-linked TmG as release retardants. The drugs (DS and ketoprofen) release from these matrix tablets expressed a sustained releasing pattern of drugs over a period of 8 hours. The drug-releasing rates from the tablets containing cross-linked TmG were slower than that of without cross-linking. In an investigation by Kulkarni et al., (2013), interpenetrating polymer networking (IPN) hydrogel tablets of TmG and sodium alginate (SA) were formulated for controlled releasing of propranolol. The IPN hydrogel tablets containing propranolol and propranolol-resin complex (resonate) were formulated by the wet granulation-covalent cross-linking method. These tablets showed drug release for up to 24 hours. In another study, Jana et al., (2014) formulated IPN matrix tablets by direct compression of aceclofenac-loaded chitosan-TmG IPN microparticles. These tablets presented sustained releasing of aceclofenac over 8 hours in the *in vitro* dissolution study, which was also found comparable with marketed commercial tablets.

1.7   ORAL MULTIPLE-UNIT SYSTEMS

TmG has been employed in the formulations of diverse particulate systems such as nanoparticles, microparticles, spheroids, beads, etc. for oral application (Nayak, 2016). The multiple-unit systems are capable of the mixture with gastrointestinal (GI) fluid and circulating in the gastrointestinal tract (GIT) over a longer area, which facilitates more expectable drug release

TABLE 1.1 Use of TmG in Tablets as Binder, Matrix Former and Release Retardant

Tablets	Applications as excipient	References
Ibuprofen tablets	Binder, Release-retardant	Kulkarni et al., 2011
Tramadol HCl tablets	Binder	Phani Kumar et al., 2011
Diclofenac sodium tablets	Binder	Kampanart et al., 2016
Acyclovir matrix tablets	Release-retardant, Matrix-former	Chandramouli et al., 2012
Aceclofenac matrix tablets	Release-retardant, Matrix-former	Basavaraj et al., 2011
Diclofenac sodium matrix tablets	Release-retardant, Matrix-former	Malviya et al., 2010; Mahavarkar et al., 2016
Lamivudine matrix tablets	Release-retardant, Matrix-former	Alka et al., 2011
Ketoprofen matrix tablets	Release-retardant, Matrix-former	Srinivasan et al., 2011
Propranolol HCl matrix tablet	Release-retardant, Matrix-former	Rathi et al., 2013
Aceclofenac matrix tablets	Release-retardant, Matrix-former	Radhika et al., 2011
Clarithromycin matrix tablets	Release-retardant, Matrix-former	Km and Arul, 2016
Salbutamol sulfate mucoadhesive sustained-release tablets	Matrix-former, Release-retardant, Mucoadhesive	Mitra et al., 2012
Terbutaline sulfate mucoadhesive sustained-release tablets	Matrix-former, Release-retardant, Mucoadhesive	Chanda and Roy, 2010
Verapamil HCl floating-bioadhesive tablets	Matrix-former, Release-retardant, Mucoadhesive	Mali and Dhawale, 2016

(Malakar and Nayak, 2012). Furthermore, multiple-unit systems obviate different GI transit rates and the vagaries of gastric emptying, drug release at the higher concentrations, when likened with the single-unit dosage forms. Multiple-unit systems also decrease the possibility of dose variation and mucosal damage (Malakar et al., 2012).

1.7.1 SPHEROIDS

DS containing spheroids was formulated using TmG by extrusion-spheronization technique (Kulkarni et al., 2005). These spheroids exhibited 8 hours release profile, *in vitro* with a controlled releasing pattern. In an aspect of *in vitro* drug releasing pattern of these TmG spheroids, the viscosity of TmG, and the swelling index were studied and analyzed.

1.7.2 CONTROLLED RELEASE MICROPARTICLES/BEADS

Novel kinds of pH-responsive beads composed of TmG-alginate for the controlled releasing of DS were formulated via an ionotropic-gelation procedure and evaluated (Nayak and Pal, 2011). These TmG-alginate beads were of 1.33 ± 0.04 mm to 0.71 ± 0.03 mm in size. The drug encapsulation efficiency was 72.23 ± 2.14 to $97.32 \pm 4.03\%$ with sustained *in vitro* DS releasing of 69.08 ± 2.36 to $96.07 \pm 3.54\%$ after 10 hours. The *in vitro* DS releasing from these TmG-alginate beads followed the controlled releasing pattern, showing the zero-order kinetics with the release mechanism of case-II transport. The degradation and swelling of the developed beads were demonstrated to be tempted by several pHs of the test intermediate.

1.7.3 INTERPENETRATING POLYMER NETWORKING (IPN) MICROPARTICLES

Kulkarni et al., (2012) formulated diltiazem-Indion 254® complex microbeads prepared using TmG and SA through ionotropic gelation as well as covalent cross-linking technique. These IPN microbeads were of 986–1257 mm in size and demonstrated 78.15–92.15% of drug entrapment efficiency. The *in vitro* DS releasing from these IPN microbeads was found to be sustained over 9 hours. On the other hand, diltiazem-Indion 254® complex showed sustained releasing 4 hours. The *in vivo* study of TmG-alginate IPN microbeads in Wister rats demonstrated more *AUC* values importantly higher bioavailability of diltiazem.

Jana et al., (2013) formulated aceclofenac-loaded chitosan-TmG IPN microparticles via a covalent cross-linking process by using glutaraldehyde (GA) at the pH of 5.5. The particle sizes were found 490.55 ± 23.24 to 621.60 ± 53.57 μm. Scanning electron microscopy (SEM) indicated an approximately spherically shaped microparticle without agglomeration. The drug entrapment efficiency was observed within 85.84 ± 1.75 to 91.97 ± 1.30%. *In vitro* releasing of aceclofenac from the IPN microparticles containing aceclofenac was estimated via the dialysis bag diffusion procedure in the phosphate buffer of pH 6.8, which demonstrated a sustained aceclofenac releasing over 8 hours. The *in vitro* releasing of the encapsulated drug demonstrated to follow the Korsmeyer-Peppas model with anomalous diffusion mechanism (non-Fickian). These IPN-based microparticulate systems demonstrated a sustained *in vivo anti-inflammatory action in the* carrageenan-induced rats *over a long time* after the oral intake.

1.7.4 MUCOADHESIVE MICROPARTICLES/BEADS

Recently, TmG was employed as mucoadhesive polymer blends to develop mucoadhesive microparticles and beads (Nayak and Pal, 2013; Nayak et al., 2014c, d). Mucoadhesive beads comprised of metformin HCl with low methoxy pectin (LMP)-TmG polymer-blends and prepared through the ionotropic-gelation procedure (Nayak et al., 2014c). The drug encapsulation efficiency of 95.12 ± 4.26%, and the mean diameter of 1.93 ± 0.26 mm were measured for the optimized beads. The SEM image suggested that these beads were of spherically shaped with the harsh surface. The *in vitro* drug release was 46.53 ± 3.28% at 10 hours. Another metformin HCl-loaded TmG-gellan gum mucoadhesive beads by the ionotropic-gelation procedure were developed by the same research group (Nayak et al., 2014d). The average diameter of optimized beads was 1.70 ± 0.24 mm, and the drug encapsulation efficiency was 95.73 ± 4.02%. The SEM photograph indicated irregular surface consisting of cracks and wrinkles in spherical beads. The *in vitro* drug release was 61.22 ± 3.44% at 10 hours. Both the metformin HCl-loaded polymeric beads (TmG-pectinate beads and TmG-gellan gum beads) pursued the controlled releasing profiles with zero-order kinetics modeling of drug-releasing with and the mechanism of super case-II transport of drug release. Both these beads in the intestinal mucosa showed high biomucoadhesivity demonstrating a faster wash off behavior in the intestinal pH. The ionization of carboxyl groups of the matrix is responsible for this biomucoadhesion phenomenon, which reduced the adhesive strength by increasing their solubility. Both the mucoadhesive beads of metformin HCl demonstrated the hypoglycemic action in the alloxan-induced diabetic rats over a long period after the oral intake, significantly.

The same research group has also developed TmG-alginate mucoadhesive beads and microspheres via the ionotropic-gelation process (Nayak and Pal, 2013; Nayak et al., 2016). Mucoadhesive beads comprised of TmG-alginate blends with metformin HCl, and its oral delivery was developed (Nayak and Pal, 2013). The finalized beads showed 94.86 ± 3.92% metformin HCl encapsulation efficiency with 1.24 ± 0.07 mm diameter in size. Super case-II transport mechanism with the controlled releasing profiles with zero-order kinetics modeling and the super case-II transport mechanism was observed. In another study, the same group created TmG-alginate biomucoadhesive microspheres encapsulated with gliclazide by changing the ratios of polymer-blends and calcium chloride concentrations in the cross-linking solution through the ionotropic-gelation process (Pal and Nayak, 2012). The gliclazide entrapments of these microspheres were 58.12 ± 2.42 to 82.78 ± 3.43% w/w; whereas the microparticle-sizes were measured as 752.12 ± 6.42 to 948.49 ± 20.92 µm. It exhibited 12 hours of prolonged *in vitro* gliclazide releasing. Both the metformin HCl-loaded beads and gliclazide-loaded microspheres made of TmG-alginate blends exhibited good *ex vivo* biomucoadhesive nature on the goat intestinal mucosal surface in the wash-off study. *In vivo* studies of these mucoadhesive beads and mucoadhesive microspheres in the alloxan-induced diabetic albino rats observed the important hypoglycemic effect of oral administration, significantly, and found a fix for management of non-insulin dependent diabetes mellitus with the maintenance of blood glucose level.

1.7.5 FLOATING BEADS

Currently, TmG was used in the development of floating gastroretentive beads (Nayak et al., 2013d). In these beads, low-density oil was trapped to accomplish the buoyancy for prolonged periods. Ionotropic emulsion gelation method was used for groundnut oil-entrapped TmG-alginate blend buoyant beads comprised of DS and utilized in gastroretentive drug delivery. The buoyant beads exhibited the drug encapsulation of 82.48 ± 2.34% w/w and the density of 0.88 ± 0.07 gram/cm^3. The optimized floating beads with DS demonstrated good anti-inflammatory action in the carrageenan-induced rat over a longer time after the oral intake.

1.8   BUCCAL DRUG DELIVERY

As a buccoadhesive polymeric agent, TmG has been utilized in different buccal drug delivery systems including buccal tablets, buccal films, and patches (Avachat et al., 2013; Bangle et al., 2011; Patel et al., 2009). Nifedipine buccoadhesive tablets of using TmG were formulated and evaluated for buccoadhesive delivery, which has shown a good mucoadhesivity with the goat buccal mucosa and sustained nifedipine releasing behavior (Patel et al., 2009). Moreover, TmG was compared with HPMC and Na CMC as buccoadhesive agents in these tablets. In another investigation, nitrendipine buccoadhesive tablets were formulated using TmG as mucoadhesive agent (Bangle et al., 2011). The mucoadhesivity potential of TmG in nitrendipine tablets was compared with polysaccharide isolated from *Ziziphus mauritiana,* HPMC, *and* Na CMC. The results confirmed the usefulness of TmG as a mucoadhesive agent in the formulation of buccal tablets. Biomucoadhesive buccal films of rizatriptan benzoate were prepared using tamarind seed xyloglucan as a biomucoadhesive agent with carbopol 934 P (Avachat et al., 2013). The *ex vivo* permeation of rizatriptan benzoate across porcine buccal mucosa using Franz diffusion cell exhibited sustained drug permeation over a prolonged time. In another study, epichlorohydrin cross-linked TmG biomucoadhesive buccal patches of metronidazole were prepared, which have exhibited high-quality *ex vivo* biomucoadhesivity onto the buccal mucosal membrane with the sustained drug permeation over a longer period (Jana et al., 2010). The drug permeation was found dependent on the degree of cross-linking. In another investigation by Ahuja et al., (2013) buccal patches of metronidazole using TmG and TmG-g-poly (N-vinyl 2-pyrrolidone) were prepared and evaluated. The *ex vivo* biomucoadhesion of the grafted polymer-based buccal patches showed biomucoadhesion of 9.3 hours with 80% < of metronidazole getting released while the TmG buccal patches exhibited *ex vivo* biomucoadhesion of 5 hours releasing metronidazole of 50%.

1.9 OCULAR DRUG DELIVERY

TmG was already investigated in the grooming of various ocular drug delivery systems like ocular gels and ocular nanoparticles (Gheraldi et al., 2000, 2004; Mehra et al., 2010). The mucoadhesive property and viscosity of TmG make it as a suitable excipient in various ocular formulations for enhancing the ocular residence period for various drug candidates on the cornea. A TmG based *in situ* gelling ocular dosage form of pilocarpine was developed and evaluated for its miotic potential (Mehra et al., 2010). The formulation of TmG, alginate, and chitosan (CS) was distinguished to the most vital means for sustained delivery of 80% pilocarpine in 12 hours. *In vivo*, miotic study and ocular irritation study exhibited a significant long-lasting decrease in pupil diameter of rabbits and well allowed non-irritating effect with TmG based preparation. The biomucoadhesive potential of TmG for ocular administration of hydrophobic and hydrophilic antibiotics like ofloxacin and gentamicin was investigated by Gheraldi et al., (2000). These ocular formulations were instilled in rabbit eyes. The results of the study exhibited that the aqueous humor and corneal concentration of the dose were remarkably higher than the drugs alone. The drug absorptions and drug eliminations for both the drug-containing formulations were prolonged by the use of TmG as a mucoadhesive agent. TmG was also investigated for ocular delivery of rufloxacin and ofloxacin to treat bacterial keratitis experimentally stimulated by *Pseudomonas aeruginosa* and *Staphylococcus aureus* in rabbits (Gheraldi et al., 2004). The result indicated a significant increment in intra-aqueous penetration of drugs in both infected and uninfected rabbit eyes. The use of TmG bettered the prolongation of precorneal residence time and drug accumulation by these formulations. In a probe, tropicamide-loaded TmG nanoaggregates were optimized and formulated for the ocular drug release (Dilbaghi et al., 2013). The optimized tropicamide-loaded nanoaggregates observed an important higher *ex vivo* corneal tropicamide permeations across the goat corneal membrane (isolated). The results were compared with the commercial aqueous formulation (conventional dosage form), and this demonstrated a good corneal biomucoadhesiveness of these nanoaggregates. These tropicamide-contained nanoaggregates also demonstrated good ocular tolerance as well as biocompatibility as dictated by the hen's egg test using the chorioallantoic membrane and resazurin assay on Vero cell lines. In another study, carboxymethyl-modified TmG was applied to develop ocular nanoparticles containing tropicamide (Kaur et al., 2012b). These nanoparticles were formulated by ionotropic-gelation process and optimized. The optimized tropicamide-loaded carboxymethyl TmG nanoparticles exhibited *ex vivo* corneal tropicamide permeation of tropicamide across the goat corneal membrane (isolated) in comparison with that of the tropicamide aqueous solution. The *ex vivo* ocular bioadhesion, as well as the ocular tolerance of these nanoparticles, showed their potential as an ocular delivery carrier system.

1.10 NASAL DRUG DELIVERY

A nasal drug delivery of diazepam composed of TmG as a biomucoadhesive agent was evaluated and acquired (Datta and Bandyopadhyay, 2006). The pH, viscosity, and gelling property of TmG were higher as compared to the synthetic polymers like HPMC and carbopol 934 (commonly utilized) in nasal drug delivery systems as a mucoadhesive agent. The *ex vivo* biomucoadhesivity of the formulations containing TmG using bovine nasal membrane was also calculated, and the result of this study confirmed higher biomucoadhesive strength than that of HPMC and carbopol 934. *In vitro* drug release characteristic through Franz-diffusion cell using bovine nasal membrane exhibited the effectiveness of mucoadhesive agent in these nasal formulations in comparison to that of HPMC and carbopol 934 without and with the permeation inducer.

1.11 COLON-TARGETED DRUG DELIVERY

The colon targeted drug releasing is a necessity to safeguard the drugs during its transit via the upper GIT and thus, permit the releasing of drugs in the colon region. TmG also was studied for biodegradable carrier and colon targeted drug delivery (Mishra and Khandare, 2011). Ibuprofen matrix tablets containing dissimilar concentrations of TmG were prepared by wet granulation technique for the protection of drugs in the upper GI portion. These tablets discharged the major amount of ibuprofen, *in vitro* in assuming the gastric fluid (pH 1.2), imitated intestinal fluid (pH 7.4) and assumed colonic fluid (2 and 4% w/v rat caecal contents, pH 6.8) before and after the enzyme induction process. The *in vitro* biodegradation evaluation results demonstrated that TmG was devalued in the comportment of the rat caecal contents.

1.12 CONCLUSION

The main objective for searching a new excipient is to overwhelm the shortcomings of preparing cost, availability, toxicity, and compatibility. Currently, TmG is attaining popularity for its utility in the preparation of numerous pharmaceutical dosage forms. The current chapter observes the chances by using TmG as potential pharmaceutical excipients in diverse pharmaceutical formulations in the pharmaceutical industry.

KEYWORDS

- **drug delivery**
- **excipient**
- **sustained drug release**
- **tamarind gum**

REFERENCES

Ahuja, M., Kumar, S., & Kumar, A., (2013). Tamarind seed polysaccharide-g-poly (N-vinyl-2-pyrrolidone): Microwave-assisted synthesis, characterization and evaluation as mucoadhesive polymer. *Int. J. Polym. Mater. Polym. Biomater.*, *62*, 544–549.

Alka, A., Laxmi, B., & Suman, R., (2011). Formulation development and evaluation of sustained release matrix tablet of lamivudine using tamarind seed polysaccharide. *Der Pharm. Lett.*, *3*, 250–258.

Avachat, A. A., Dash, R. R., & Shrotriya, S. N., (2011). Recent investigations of plant-based natural gums, mucilages and resins in novel drug delivery systems. *Indian J. Pharm. Educ. Res.*, *45*, 86–99.

Avachat, A., Gujar, K. N., & Wagh, K. V., (2013). Development and evaluation of tamarind seed glucan-based mucoadhesive buccal films of rizatriptan benzoate. *Carbohydr. Polym.*, *91*, 537–542.

Bangle, G. S., Shinde, G. V., Umalkar, D. G., & Rajesh, K. S., (2011). Natural mucoadhesive material based buccal tablets of nitrendipine–formulation and *in-vitro* evaluation. *J. Pharm. Res.*, *4*, 33–38.

Basavaraj, B. V., Someswara, R. B., Kulkarni, S. V., Patil, P., & Surpur, C., (2011). Design and characterization of sustained release aceclofenac matrix tablets containing tamarind seed polysaccharide. *Asian J. Pharm. Tech.*, *1*, 17–21.

Bera, H., Boddupalli, S., & Nayak, A. K., (2015a). Mucoadhesive-floating zinc-pectinate-sterculia gum interpenetrating polymer network beads encapsulating ziprasidone HCl. *Carbohydr. Polym.*, *131*, 108–118.

Bera, H., Boddupalli, S., Nandikonda, S., Kumar, S., & Nayak, A. K., (2015b). Alginate gel-coated oil-entrapped alginate–tamarind gum–magnesium stearate buoyant beads of risperidone. *Int. J. Biol. Macromol.*, *78*, 102–111.

Bera, H., Kandukuri, S. G., Nayak, A. K., & Boddupalli, S., (2015c). Alginate-sterculia gum gel-coated oil-entrapped alginate beads for gastroretentive risperidone delivery. *Carbohydr. Polym.*, *120*, 74–84.

Biswas, N., & Sahoo, R. K., (2016). Tapioca starch blended alginate mucoadhesive-floating beads for intragastric delivery of metoprolol tartrate. *Int. J. Biol. Macromol.*, *83*, 61–70.

Chanda, R., & Roy, A., (2010). Formulation of terbutaline sulfate mucoadhesive sustained release oral tablets from natural materials and *in vitro-in vivo* evaluation. *Asian J. Pharm.*, *4*, 168–174.

Chandramouli, Y., Firoz, S., Vikram, A., Padmaja, C., & Chakravarthi, N. R., (2012). Design and evaluation of controlled release matrix tablets of acyclovir sodium using tamarind seed polysaccharide. *J. Pharm. Biol.*, *2*, 55–62.

Das, B., Nayak, A. K., & Nanda, U., (2013). Topical gels of lidocaine HCl using cashew gum and Carbopol 940: Preparation and *in vitro* skin permeation. *Int. J. Biol. Macromol.*, *62*, 514–517.

Deveswaran, R., Bharath, S., Furtado, S., Basavraj, B. V., Furtado, S., & Madhavan, V., (2010). Isolation and evaluation of tamarind seed polysaccharide as a natural emulsifying agent. *Int. J. Pharm. Biol. Arch.*, *1*, 360–363.

Dilbaghi, N., Kaur, H., Ahuja, M., & Kumar, S., (2013). Evaluation of tropicamide-loaded tamarind seed xyloglucan nanoaggregates for ophthalmic delivery. *Carbohydr. Polym.*, *94*, 286–291.

Gheraldi, E., Tavanti, A., Celandroni, F., Lupetti, A., Blandizzi, C., Boldrini, E., Campa, M., & Senesi, S., (2000). Effect of novel mucoadhesive polysaccharide obtained from tamarind seeds on the intraocular penetration of gentamicin and ofloxacin in rabbits. *J. Antimicrob. Chemother.*, *46*, 831–834.

Gheraldi, E., Tavanti, A., Davini, P., Celandroni, F., Salvetti, S., Parisio, E., et al., (2004). A mucoadhesive polymer extracted from tamarind seed improves the intraocular penetration and efficacy of rufloxacin in topical treatment of experimental bacterial keratitis. *Antimicrob. Agents Chemother.*, *48*, 3396–3401.

Ghosh, S., & Pal, S., (2013). Modified tamarind kernel polysaccharide: A novel matrix for control release of aspirin. *Int. J. Biol. Macromol.*, *58*, 296–300.

Guru, P. R., Bera, H., Das, M., Hasnain, M. S., & Nayak, A. K., (2018). Aceclofenac-loaded *Plantago ovata* F. husk mucilage-Zn^{+2}-pectinate controlled-release matrices. *Starch - Stärke*, *70*, 1700136.

Guru, P. R., Nayak, A. K., & Sahu, R. K., (2013). Oil-entrapped sterculia gum-alginate buoyant systems of aceclofenac: Development and *in vitro* evaluation. *Colloids Surf. B: Biointerf.*, *104*, 268–275.

Hasnain, M. S., Nayak, A. K., Singh, R., & Ahmad, F., (2010). Emerging trends of natural-based polymeric systems for drug delivery in tissue engineering applications. *Sci. J. UBU.*, *1*, 1–13.

Hasnain, M. S., Rishishwar, P., & Ali, S., (2017a). Floating-bioadhesive matrix tablets of hydralazine HCL made of cashew gum and HPMC K4M. *Int J Phar Pharm Sci., 9*(7), 124-129.

Hasnain, M. S., Rishishwar, P., & Ali, S., (2017b). Use of cashew bark exudate gum in the preparation of 4% lidocaine HCL topical gels. *Int. J. Phar. Pharm. Sci., 9*(8), 146–150.

Hasnain, M. S., Rishishwar, P., Rishishwar, S., Ali, S., & Nayak, A. K., (2018a). Extraction and characterization of cashew tree (*Anacardium occidentale*) gum, use in aceclofenac dental pastes. *Int. J. Biol. Macromol., 116*, 1074–1081.

Hasnain, M. S., Rishishwar, P., Rishishwar, S., Ali, S., & Nayak, A. K., (2018b). Isolation and characterization of *Linum usitatisimum* polysaccharide to prepare mucoadhesive beads of diclofenac sodium. *Int. J. Biol. Macromol., 116*, 162–172.

Jana, S., Lakshman, D., Sen, K. K., & Basu, S. K., (2010). Development and evaluation of epichlorohydrin cross-linked mucoadhesive patches of tamarind seed polysaccharide for buccal application. *Int. J. Pharm. Sci. Drug. Res., 2*, 193–198.

Jana, S., Saha, A., Nayak, A. K., Sen, K. K., & Basu, S. K., (2013). Aceclofenac-loaded chitosan-tamarind seed polysaccharide interpenetrating polymeric network microparticles. *Colloid. Surf. B: Biointerf., 105*, 303–309.

Jana, S., Sen, K. K., & Basu, S. K., (2014). *In vitro* aceclofenac release from IPN matrix tablets composed of chitosan-tamarind seed polysaccharide. *Int. J. Biol. Macromol., 65*, 241–245.

Jena, A. K., Nayak, A. K., De, A., Mitra, D., & Samanta, A., (2018). Development of lamivudine containing multiple emulsions stabilized by gum odina. *Future J. Pharm. Sci., 4*, 71–79.

Joseph, J., Kanchalochana, S. N., Rajalakshmi, G., Hari, V., & Durai, R. D., (2012). Tamarind seed polysaccharide: A promising natural excipients for pharmaceuticals. *Int. J. Green Pharm., 6*, 270–278.

Kampanart, H., Tanikan, S., & Wancheng, S., (2016). Development of tamarind seed gum as dry binder in formulation of diclofenac sodium tablets. *Walailak J. Sci. Technol., 13*, 863–874.

Kaur, H., Ahuja, M., Kumar, S., & Dilbaghi, N., (2012b). Carboxymethyl tamarind kernel polysaccharide nanoparticles for ophthalmicdrug delivery. *Int. J. Biol. Macromol., 50*, 833– 839.

Kaur, H., Yadav, S., Ahuja, M., & Dilbaghi, N., (2012a). Synthesis, characterization and evaluation of thiolated tamarind seed polysaccharide as a mucoadhesive polymer. *Carbohydr. Polym., 90*, 1543–1549.

Km, N., & Arul, B., (2016). Design and evaluation of clarithromycin sustained release tablets using isolated natural polysaccharide. *Indo Am. Journal of Pharm. Res., 6*, 6469–6476.

Kulkarni, G. T., Gowthamarajan, K., Dhobe, R. R., Yohanan, F., & Suresh, B., (2005). Development of controlled release spheroids using natural polysaccharide as release modifier. *Drug Deliv., 12*, 201–206.

Kulkarni, G. T., Seshubabu, P., & Kumar, S. M., (2011). Effect of tamarind seed polysaccharide on dissolution behavior of ibuprofen tablets. *J. Chronother. Drug Deliv., 2*, 49–56.

Kulkarni, R. V., Mutalik, S., Mangond, B. S., & Nayak, U. Y., (2012). Novel interpenetrated polymer network microbeads of natural polysaccharides for modified release of water-soluble drug: *in vitro* and *in vivo* evaluation. *J. Pharm. Pharmacol., 64*, 530–540.

Kumar, R., Patil, S. R., Patil, M. B., Paschapur, M. S., & Mahalaxmi, R., (2001). Isolation and evaluation of tamarind seed polysaccharide on castor oil emulsion. *Der Pharm. Lett.*, *2*, 518–521.

Mahavarkar, R. V., Ahirrao, S., Kshirsagar, S., & Rayate, V., (2016). Formulation and evaluation of tamarind seed polysaccharide matrix tablet. *Pharm. Biol. Eva.*, *3*, 241–255.

Malakar, J., & Nayak, A. K., (2012). Formulation and statistical optimization of multiple-unit ibuprofen-loaded buoyant system using 2^3-factorial design. *Chem. Eng. Res. Des.*, *9*, 1834–1846.

Malakar, J., Nayak, A. K., & Pal, D., (2012). Development of cloxacillin loaded multiple-unit alginate-based floating system by emulsion–gelation method. *Int. J. Biol. Macromol.*, *50*, 138–147.

Malakar, J., Nayak, A. K., Jana, P., & Pal, D., (2013). Potato starch-blended alginate beads for prolonged release of tolbutamide: Development by statistical optimization and *in vitro* characterization. *Asian J. Pharm.*, *7*, 43–51.

Malaviya, R., Srivastava, P., Kumar, U., Bhargava, C. S., & Sharma, P. K., (2010). Formulation and comparison of suspending properties of different natural polymers using paracetamol suspension. *Int. J. Drug Dev. Res.*, *2*, 886–891.

Mali, K. K., & Dhawale, S. C., (2016). Design and optimization of modified tamarind gum-based floating-bioadhesive tablets of verapamil hydrochloride. *Asian J. Pharm.*, *10*, 239–250.

Malviya, R., Srivastava, P., Bansal, V., & Sharma, P. K., (2010). Formulation evaluation and comparison of sustained release matrix tablets of diclofenac sodium using natural polymers as release modifier. *Int. J. Pharm. Biol. Sci.*, *1*, 1–8.

Manchanda, R., Arora, S. C., & Manchanda, R., (2014). Tamarind seed polysaccharide and its modification-versatile pharmaceutical excipients-A review. *Int. J. Pharm. Tech Res.*, *6*, 412–420.

Meenakshi, & Ahuja, M., (2015). Metronidazole loaded carboxymethyl tamarind kernel polysaccharide-polyvinyl alcohol cryogels: Preparation and characterization. *Int. J. Biol. Macromol.*, *72*, 931–938.

Mehra, G. R., Manish, M., Rashi, S., Neeraj, G., & Mishra, D. N., (2010). Enhancement of miotic potential of pilocarpine by tamarind gum based *in situ* gelling ocular dosage form. *Acta Pharm. Sci.*, *52*, 145–154.

Mishra, M. U., & Khandare, J. N., (2011). Evaluation of tamarind seed polysaccharide as a biodegradable carrier for colon-specific drug delivery. *Int. J. Pharm. Pharmaceut. Sci.*, *3*, 139–142.

Mitra, T., Pattnaik, S., Panda, J., Rout, B. K., Murthy, P. S. R., & Sahu, R. K., (2012). Formulation and evaluation of salbutamol sulfate mucoadhesive sustained release tablets using natural excipients. *J. Chem. Pharm. Sci.*, *5*, 62–66.

Nandi, R. C., (1975). A process for preparation of polyose from the seeds of *Tamarindus indica*: *Indian Patent 142092*.

Nayak, A. K., (2016). Tamarind seed polysaccharide-based multiple-unit systems for sustained drug release. In: Kalia, S., & Averous, L., (eds.), *Biodegradable and Bio-Based Polymers: Environmental and Biomedical Applications* (pp. 469–492). Wiley-Scrivener, USA.

Nayak, A. K., & Pal, D., (2011). Development of pH-sensitive tamarind seed polysaccharide-alginate composite beads for controlled diclofenac sodium delivery using response surface methodology. *Int. J. Biol. Macromol., 49,* 784–793.

Nayak, A. K., & Pal, D., (2013). Ionotropically-gelled mucoadhesive beads for oral metformin HCl delivery: Formulation, optimization and antidiabetic evaluation. *J. Sci. Ind. Res., 72,* 15–22.

Nayak, A. K., & Pal, D., (2015). Chitosan-based interpenetrating polymeric network systems for sustained drug release. In: Tiwari, A., Patra, H. K., & Choi, J. W., (eds.), *Advanced Theranostics Materials* (pp. 183–208). Wiley-Scrivener, USA.

Nayak, A. K., & Pal, D., (2016a). Sterculia gum-based hydrogels for drug delivery applications. In: Kalia, S., (ed.), *Polymeric Hydrogels as Smart Biomaterials, Springer Series on Polymer and Composite Materials* (pp. 105–151). Springer International Publishing, Switzerland.

Nayak, A. K., & Pal, D., (2016b). Plant-derived polymers: Ionically gelled sustained drug release systems, In: Mishra, M., (ed.), *Encyclopedia of Biomedical Polymers and Polymeric Biomaterials* (Vol. VIII, pp. 6002–6017). Taylor & Francis Group, New York, NY 10017, USA.

Nayak, A. K., Ara, T. J., Hasnain, M. S., & Hoda, N., (2018). Okra gum-alginate composites for controlled releasing drug delivery, In: Inamuddin, A. A. M., & Mohammad, A., (eds.), *Applications of Nanocomposite Materials in Drug Delivery* (pp. 761–785). A volume in Woodhead Publishing Series in Biomaterials, Elsevier Inc.

Nayak, A. K., Beg, S., Hasnain, M. S., Malakar, J., & Pal, D., (2018). Soluble starch-blended Ca^{2+}-Zn^{2+}-alginate composites-based microparticles of aceclofenac: Formulation development and *in vitro* characterization. *Future J. Pharm. Sci., 4,* 63–70.

Nayak, A. K., Hasnain, M. S., & Pal, D., (2018). Gelled microparticles/beads of sterculia gum and tamarind gum for sustained drug release, In: Thakur, V. K., & Thakur, M. K., (eds.), *Handbook of Springer on Polymeric Gel* (pp. 361–414). Springer International Publishing, Switzerland.

Nayak, A. K., Pal, D. (2017). Natural starches-blended ionotropically gelled micropar-ticles/beads for sustained drug release, In: Thakur, V. K., Thakur, M. K., & Kessler, M. R., (eds.), *Handbook of Composites from Renewable Materials* (Vol. 8, pp. 527–560). Wiley-Scrivener, USA.

Nayak, A. K., Pal, D., & Das, S., (2013a). Calcium pectinate-fenugreek seed mucilage mucoadhesive beads for controlled delivery of metformin HCl. *Carbohydr. Polym., 96,* 349–357.

Nayak, A. K., Pal, D., & Hasnain, M. S., (2013b). Development and optimization of jackfruit seed starch-alginate beads containing pioglitazone. *Curr. Drug Deliv., 10,* 608–619.

Nayak, A. K., Pal, D., & Malakar, J., (2013d). Development, optimization and evaluation of emulsion-gelled floating beads using natural polysaccharide-blend for controlled drug release. *Polym. Eng. Sci., 53,* 338–350.

Nayak, A. K., Pal, D., & Santra, K., (2013c). *Plantago ovata* F. Mucilage-alginate mucoadhesive beads for controlled release of glibenclamide: Development, optimization, and *in vitro-in vivo* evaluation. *J. Pharm.,* Article ID 151035.

Nayak, A. K., Pal, D., & Santra, K., (2014a). *Artocarpus heterophyllus* L. seed starch-blended gellan gum mucoadhesive beads of metformin HCl. *Int. J. Biol. Macromol., 65,* 329–339.

Nayak, A. K., Pal, D., & Santra, K., (2014b). Ispaghula mucilage-gellan mucoadhesive beads of metformin HCl: Development by response surface methodology. *Carbohydr. Polym.*, *107*, 41–50.

Nayak, A. K., Pal, D., & Santra, K., (2014c). Development of calcium pectinate-tamarind seed polysaccharide mucoadhesive beads containing metformin HCl. *Carbohydr. Polym.*, *101*, 220–230.

Nayak, A. K., Pal, D., & Santra, K., (2014d). Tamarind seed polysaccharide-gellan mucoadhesive beads for controlled release of metformin HCl. *Carbohydr. Polym.*, *103*, 154–163.

Nayak, A. K., Pal, D., & Santra, K., (2015). Screening of polysaccharides from tamarind, fenugreek and jackfruit seeds as pharmaceutical excipients. *Int. J. Biol. Macromol.*, *79*, 756–760.

Nayak, A. K., Pal, D., & Santra, K., (2016). Swelling and drug release behavior of metformin HCl-loaded tamarind seed polysaccharide-alginate beads. *Int. J. Biol. Macromol.*, *82*, 1023–1027.

Nayak, A. K., Pal, D., Pany, D. R., & Mohanty, B., (2010). Evaluation of *Spinacia oleracea* L. leaves mucilage as innovative suspending agent. *J. Adv. Pharm. Technol. Res.*, *1*, 338–344.

Nayak, A. K., Pal, D., Pradhan, J., & Ghorai, T., (2012). The potential of *Trigonella foenum-graecum* L. seed mucilage as suspending agent. *Indian J. Pharm. Educ. Res.*, *46*, 312–317.

Pal, D., & Nayak, A. K., (2012). Novel tamarind seed polysaccharide-alginate mucoadhesive microspheres for oral gliclazide delivery. *Drug Deliv.*, *19*, 123–131.

Pal, D., & Nayak, A. K., (2015). Alginates, blends and microspheres: Controlled drug delivery, In: Mishra, M., (ed.), *Encyclopedia of Biomedical Polymers and Polymeric Biomaterials* (Vol. I, pp. 89–98). Taylor & Francis Group, New York, NY 10017, USA.

Pal, D., & Nayak, A. K., (2017). Plant polysaccharides-blended ionotropically-gelled alginate multiple-unit systems for sustained drug release, In: Thakur, V. K., Thakur, M. K., & Kessler, M. R., (eds.), *Handbook of Composites from Renewable Materials* (Vol. 6, pp. 399–400). Wiley-Scrivener, USA.

Pal, D., Nayak, A. K., & Kalia, S., (2010). Studies on *Basella alba* L. leaves mucilage: Evaluation of suspending properties. *Int. J. Drug Discov. Technol.*, *1*, 15–20.

Patel, B., Patel, P., Bhosale, A., Hardikar, S., Mutha, S., & Chaulang, G., (2009). Evaluation of tamarind seed polysaccharide (TSP) as a mucoadhesive and sustained release component of nifedipine buccoadhesive tablet and comparison with HPMC and Na CMC. *Int. J. Pharm. Tech Res.*, *1*, 404–410.

Phani, K. G. K., Battu, G., & Raju, L., (2011). Studies on the applicability of tamarind seed polymer for the design of sustained release matrix tablet of tramadol HCl. *J. Pharm. Res.*, *4*, 703–705.

Prajapati, V. D., Jani, G. K., Moradiya, N. G., & Randeria, N. P., (2013). Pharmaceutical applications of various natural gums, mucilages and their modified forms. *Carbohydr. Polym.*, *92*, 1685–1699.

Radhika, P. R., Kharkate, P. R., & Sivakumar, T., (2011). Formulation of aceclofenac sustained release matrix tablet using hydrophilic natural gum. *Int. J. Res. Appl. Pharm.*, *2*, 851–857.

Rajab, M., Tounsi, A., Jouma, M., Neubert, R. H., & Dittgen, M., (2012). Influence of tamarind seed gum derivatives on the *in vitro* performance of gastro-retentive tablets based on hydroxypropyl methylcellulose. *Pharmazie, 67,* 956–957.

Rao, P. S., & Srivastava, H. C., (1973). Tamarind. In: Whistler, R. L., (ed.), *Industrial Gums* (2nd edn., pp. 369–411) New York: Academic Press.

Rao, P. S., Ghosh, T. P., & Krishna, S., (1946). Extraction and purification of tamarind seed polysaccharide. *J. Sci. Ind. Res., 4,* 705.

Rathi, R. C., Bundela, M. N., Patil, K. S., & Yeole, P. G., (2013). Design and characterization of propranolol HCl tablet containing tamarind seed polysaccharide as a release retardant. *Int. J. Inst. Pharm. Life Sci., 3,* 1–12.

Samal, P. K., & Dang, J. S., (2014). Isolation, preliminary characterization and hepato-protective activity of polysaccharides from *Tamarindus indica* L. *Carbohydr. Polym., 102,* 1–7.

Shaikh, S. S., Shivsaran, K. J., Pawar, R. K., Misal, N. S., Mene, H. R., & More, B. A., (2015). Tamarind seed polysaccharide: A versatile pharmaceutical excipient and its modification. *Int. J. Pharm. Sci. Rev. Res., 33,* 157–164.

Sinha, P., Ubaidulla, U., & Nayak, A. K., (2015b). Okra (*Hibiscus esculentus*) gum-alginate blend mucoadhesive beads for controlled glibenclamide release. *Int. J. Biol. Macromol., 72,* 1069–1075.

Sinha, P., Ubaidulla, U., Hasnain, M. S., Nayak, A. K., & Rama, B., (2015a). Alginate-okra gum blend beads of diclofenac sodium from aqueous template using $ZnSO_4$ as a cross-linker. *Int. J. Biol. Macromol., 79,* 555–563.

Sravani, B., Deveswara, R., Bharath, S., Basavraj, B. V., & Mahadevan, V., (2011). Release characteristics of drugs from cross-linked tamarind seed polysaccharide matrix tablets. *Der Pharm. Lett., 2,* 67–76.

Srinivasan, B., Ganta, A., Rajamanickam, D., Veerabhadraiah, B. B., & Varadharajan, M., (2011). Evaluation of tamarind seed polysaccharide as a drug release retardant. *Int. J. Pharm. Sci. Res. Rev., 9,* 27–31.

Tattiyakul, J., Muangnapoh, C., & Poommarinvarakul, S., (2010). Isolation and rheological properties of tamarind seed polysaccharide from tamarind kernel powder using protease enzyme and high-intensity ultrasound. *J. Food Sci., 75,* 253–260.

CHAPTER 2

Pharmaceutical Applications of Gum Arabic

JIAN GUAN and SHIRUI MAO

School of Pharmacy, Shenyang Pharmaceutical University, 103, Wenhua Road, Shenyang, 110016, China

ABSTRACT

Gum Arabic or gum acacia (GA) is an exudate from Acacia seyal and Acacia senegal tree and existed as complex. This dried sap is a global commodity with immense commercial value and generally harvested in Western Asia and Africa. Generally, it is recognized safely and popular in both food and pharmaceutical industries owing to its low solution viscosity, non-digestibility, biocompatibility, and biodegradable properties. The objective of this chapter is to make an overall introduction of GA and present the main applications of GA in the pharmaceutical field. First of all, the sources and production of GA, structure, properties, and biological effect of GA were described. Subsequently, the applications of GA in the drug delivery system were described. GA has been widely employed as an emulsifier, stabilizer, and suspending agent in colloidal drug delivery system, as an excipient in tablets, as the wall forming the material of microcapsules, microspheres, nanoparticles, and coating material of nanoparticles. Besides, GA could also be used as a hydrogel ingredient, as an additive in films; modified GA conjugate could be designed for specific drug delivery. Meanwhile, some other applications of GA were summarized in this chapter, including as a shelf-life enhancer, as well as a tumor imaging agent. In conclusion, GA is a promising ingredient for drug delivery system design, and further exploration of its usage in the pharmaceutical field is essential in the near future.

2.1   INTRODUCTION

In recent years, polysaccharides have been extensively applied in the food industry, cosmetics, and pharmaceutical fields (Coviello et al., 2007). There are many literatures related to the application of polysaccharides in pharmaceutics including chitosan (CS), alginate, and carrageenan (Agüero et al., 2017; Li et al., 2014; Mao et al., 2010; Mirtič et al., 2018). CS, alginate, and carrageenan have been extensively used in the formulation of polyelectrolyte complexes, microparticles and hydrogels to delivery both small molecular drugs and biomacromolecules owing to their physicochemical properties and mild gelation condition (Belbekhouche et al., 2018; Huang et al., 2018; Lasareva et al., 2018; Li et al., 2014; Mirtič et al., 2018). Recently, due to the good biocompatibility, biodegradability, non-toxicity, and several specific properties attributed from its structure, GA has become a promising pharmaceutical excipient and is attracting more and more attention (Verbeken et al., 2003).

As an old and well-known polysaccharide gum, the application of GA could be traced back to 5000 years ago. It is extracted from either Acacia Senegal or Acacia seyal trees (Patel and Goyal, 2015). GA is a complex polysaccharide with branches, presenting either slightly acidic or neutral properties. The backbone of GA consisted of 1, 3-linked β-D-galactopyranosyl and the side chains, which was compromised of two to five 1,3-linked β-D-galactopyranosyl was connected to the main chain via 1,6-linkage. Meanwhile, β-D-glucuronopyranosyl, α-l-arabinofuranosyl, 4-O-methyl-β-D-glucuronopyranosyl and α-l-rhamnopyranosyl, as well as some proteins, also exist along with the molecules.

GA has been extensively employed in the food industry due to the high water solubility, safety, edibility, and lack of aftertaste (Patel and Goyal, 2015). It could be used as thickener, stabilizer and emulsifying agent in ice cream, jellies, candies, soft drinks, syrups, chewing gums, and beverages as well as confectionery coatings and glazes (Pizzoni et al., 2015). Furthermore, in the Middle Eastern countries, it is regarded as a therapeutical agent for chronic diseases of the kidney (Nasir et al., 2012).

Recently, GA is increasingly applied in various non-food industries, especially in the pharmaceutical field. GA has been proved to be an important emulsifier and stabilizer for emulsions and colloids such as microparticles and nanoparticles, making GA an excellent ingredient for pharmaceutical application (Akhtar and Dickinson, 2007; Leroux et al.,

2003). It has also been used in the preparation of microcapsules, microspheres, and hydrogels (Ji et al., 2011; Krishnan et al., 2005; Paulino et al., 2006; Weinbreck et al., 2003). However, GA is less frequently reported in drug delivery systems compared with the other commonly used polymers such as CS, Carbomer, alginate (SA) and hydroxypropyl methylcellulose (HPMC). Systemic summary of the application of GA in drug delivery is still scarce. Therefore, the objective of this chapter is to present the basic properties of GA summarize its applications in the pharmaceutical field from different aspects.

2.2 GENERAL PROPERTIES OF GA

2.2.1 THE SOURCES OF GA

GA is a dried and edible gummy exudate contributes from Acacia Senegal and Acaciaseyal's branches and stems. The non-viscous soluble fibers are abundant, which belongs to the Leguminosae family (Ali et al., 2009; Patel and Goyal, 2015). In Africa, the most important and famous tree species for the production of GA is Acacia senegal (L.) Willdenow and tapped by the local populations (He et al., 2015). In the world, approximately 60% of GA was produced from Sudan, followed by 24% and 6% coming from Chad and Nigeria respectively, also, another 8% from other countries with minor production scale. GA is generally recognized as safe (GRAS) polysaccharide owing to their non-toxicity, good biocompatibility, and biodegradability.

2.2.2 THE STRUCTURE AND PROPERTIES OF GA

GA is a complex polysaccharide with branched-chain exhibiting either slightly acidic or neutral properties and existed as mixed magnesium, calcium, and potassium salt. The backbone of GA consisted of 1,3-linked β-D-galactopyranosyl and the side chains, which was compromised of two to five 1,3-linked β-D-galactopyranosyl connected to the main chain via 1,6-linkage. The units of α-L-rhamnopyranosyl, α-L-arabinofuranosyl are existed in the molecule, with β-D-glucuronopyranosyl and 4-O-methyl-b-D-glucuronopyranosyl contained as end units (Ali et al., 2009; Verbeken et al., 2003). As a complex polysaccharide, the composition of GA varied with climate, soil environment, the age, and the source of the trees picked (Assaf et al., 2005). There are approximately 12–16% rhamnose, 15–16% glucuronic acid, 24–27% arabinose, 39–42% galactose, and 12.5–16.0% moisture. Meanwhile, there are also 1.5–2.6% protein and 0.22–0.39% nitrogen presented in GA (Ali et al., 2009). Also, GA is a highly heterogeneous polysaccharide and can generally be divided into three major portions. The first part is comprised of arabinogalactan with extremely low content of protein (0.35%) with a 3.8×10^5 Da molecular weight, which made up of 88.4% of the total gum. The second fraction consisted of an arabinogalactan-protein complex with 11.8% protein of molecular weight 1.45×10^6 Da contained, which comprised of 10.4% of the gum. The arabinogalactan–protein complex (GAGP), in which the arabinogalactan chains are linked to protein chain via serine and hydroxyproline groups covalently, plays an important role in both structure and properties of GA even though with a minimal component. Although there is no clear investigation on the conformation and action at the interface, it has been considered that the emulsifying and stabilizing properties of GA are mainly attributed to the GAGP (Dror et al., 2006; Phillips, 1998). The third fraction, which contributes to 1.2% of GA refers to a low molecular weight glycoprotein, which possesses 47.3% protein content with molecular weight 2.5×10^5 Da. Above all, the most commonly contained amino acids of GA are hydroxyproline, serine, proline, and aspartic acid.

2.2.3 BIOLOGICAL PROPERTIES OF GA

2.2.3.1   EFFECT OF GA ON RENAL FUNCTION

Generally, either dialysis or renal transplantation was needed for the end stage of renal disease and renal failure. Most protein restriction regimen dietary attempts have been made to treat chronic renal failure (CRF) (Chaturvedi and Jones, 2007). Meanwhile, supplementation of the diet by fermentable carbohydrate (FC) has been proposed recently as an alternative dietetic approach (Winchester and Salsberg, 2004). Recently, it was reported that GA could not only increase fecal weight consistency but also bind with free water, resulting in absorption reduction of intestinal fluid, therefore, a lower urine volume. Furthermore, GA could also bind with Na^+ in the intestine, reducing renal excretion (Ali et al., 2009). Nasir et al., (2008) reported that GA increased the 24 h clearance of creatinine and excretion of urinary antidiuretic hormone, meanwhile, and the daily urine output, as well as the urinary Na^+ excretion, were decreased. Collectively, when treated with GA, a significantly improved creatinine clearance was observed, and the electrolyte excretion was also altered, which benefits to the renal insufficiency (Nasir et al., 2008). On the other hand, the concentration of serum butyrate was also increased by GA treatment in healthy subjects, and this may, in turn, be a salutatory effect on the clearance of creatinine (Matsumoto et al., 2006).

2.2.3.2   EFFECT OF GA ON BLOOD GLUCOSE CONCENTRATION

Wadood et al., reported that when the powdered acacia was orally administrated to normal and alloxan-induced diabetes rabbits, it could significantly reduce the concentration of blood glucose in normal rabbits, however, no decrease of blood glucose concentration was observed in diabetic rabbits (Wadood et al., 1989). Thus, the author drew the conclusion that GA could initiate the pancreatic β cells of normal rabbits to release insulin. Although there was not a clear mechanism under the reduced blood glucose concentration, the mixture of a different kind of gum was evaluated and used as food-grade viscous polysaccharides for an alternative method to reduce the postprandial hyperglycemia. It has shown that the glucose movement *in vitro* and the postprandial blood glucose could be inhibited and lowered as well as plasma insulin in human subjects by gum mixtures and this may attribute partly to the effect of viscosity on glycemic response, whereas it could slow down the emptiest of gastric (Leclere et al., 1994).

2.2.3.3   EFFECTS OF GA ON HEPATIC MACROPHAGES

It was reported that GA could almost completely block the function of macrophages in the liver (Mochida et al., 1990). This was proved by the results that GA could suppress macrophage activation *in vitro* due to their production superoxide anions (Mochida et al., 1996). Since the damage of sinusoidal endothelial cell was caused by hepatic macrophages activation induced by endotoxin, such effect of GA would be worthy of consideration in the chronic liver disease therapy, as a deranged function of hepatic macrophages and Kupffer cells occurs in this disease as well as its complications, such as endotoxemia.

2.2.3.4   EFFECT OF GA ON DENTAL PROTECTION

As well known, the acidic soft drinks may lead to human enamels softening and dissolution, which is termed as erosion (Tahmassebi et al., 2006). This could be explained by the demineralization of the enamel surface induced by citric and phosphoric acid from acidic soft drinks (Beyer et al., 2010). Interestingly, a few studies revealed that some polymers could prevent the erosion of enamel. For instance, by adding GA into drinks, the in-vitro erosion extent of enamel was decreased (Beyer et al., 2010). This could occur due to the enamel surface adsorption of GA to form a protective polymer layer. Furthermore, GA was found to have the capacity to re-mineral the enamel (Onishi et al., 2008). When demineralization and remineralization were imbalanced, GA could enhance the remineralization due to the high Ca^{2+} concentration of this polymer. This was proved by the fact that the molar remineralization ratio was significantly higher when exposed to GA than that of distilled water and the effect of GA was similar to that of NaF (Onishi et al., 2008). The GA also has the ability to inhibit early dental plaque deposition in chewing gum (Pradeep et al., 2010).

2.2.3.5   EFFECT OF GA ON THE BLOOD LIPID METABOLISM

It was demonstrated that GA could decrease 6% and 10.4% total serum cholesterol when orally administrated at either 25 g/day or30 g/day for periods of 21 or 30 days respectively (Ross et al., 1983) and it was confirmed that only the decrease of cholesterol, very low-density lipoprotein (VLDL) and low-density lipoprotein (LDL) were observed but not for high-density lipoprotein (HDL) and triglycerides. GA did not affect the plasma cholesterol concentration, but a significant lower of plasma triacylglycerol was observed (Annison et al., 1995). However, the effect of GA on the metabolism of lipid is quite variable. GA was revealed to have no significant effect on cholesterol concentration either in plasma or in the liver, and the concentration of triacylglycerol in plasma was, however, higher when fed with GA (Jensen et al., 1993). The most accepted mechanism of the cholesterol-lowering effect was related to the fecal bile acid increase and the excretion of neutral sterol as well as modification of absorption and lipid digestion (Eastwood, 1992; Moundras et al., 1994). Generally, the bile acids could be bonded or sequestered by dietary fibers, and their active reabsorption in the ileum could be diminished, resulting in excretion in feces. Consequently, the plasma cholesterol concentration decreased (Pasquier et al., 1996). Although there are various mechanisms to demonstrate the GA's effect on lipid metabolism, no clear mechanism is available, especially for the unaffected plasma cholesterol by GA, and the deep mechanism should be further explored.

2.2.3.6 OTHER BIOLOGICAL PROPERTIES

GA, as a favorable polysaccharide, has been widely employed in either the food industry or pharmaceutical field. In addition to the biological properties mentioned above, GA has been reported to have an anti-obese effect (Ushida et al., 2011), antimicrobial activity (Nishi et al., 2007b), anti-inflammatory, and anticoagulation effect (Abd El-Mawla and Osman, 2011).

2.3 APPLICATIONS OF GA IN DRUG DELIVERY

2.3.1 GA AS AN EMULSIFYING AGENT, STABILIZER, AND SUSPENDING AGENT

The emulsion-based delivery system has a wide application in the formulation of pharmaceuticals, food, and cosmetics, which can provide protection to the active ingredients against chemical degradation, with controlled release and enhanced bioavailability (McClements, 2012). It's a system comprised of two liquids that are immiscible, one termed as the internal or dispersed phase and dispersed into another phase named as external or continuous phase (Jafari et al., 2008). However, the emulsion is a highly unstable system from a thermodynamic point of view and tends to breakdown or phase separate into their individual phases during storage. Thus, emulsifier and thickening agents are usually required for improving the stability of emulsions.

GA could be applied as an emulsifier or stabilizer in emulsions attributed to its GAGP portion with the surfactant-free emulsification effect. The emulsifying effect of GA was found to be concentration dependent and mainly attributed to the GAGP, which plays a key role in its surfactant property. Furthermore, the electrostatic repulsion and steric hindrance provided by surface adsorption of GA onto micro/nanoparticles made it an excellent stabilizer for the colloidal systems (Dror et al., 2006; Phillips, 1998).

As reported, GA alone could be employed as an emulsifier, thickening or suspending agent (Liu et al., 2015) and its function was concentration dependent. For instance, GA was used as an emulsifier to prepare emulsions containing sweet almond oil at 0.5% and 2.5% concentration, respectively (Bouyer et al., 2011b). The results demonstrated that the emulsions' stability was enhanced with the increase of GA concentration, and the emulsion prepared with 0.5% GA creamed within 24 h, however, the 2.5% GA containing emulsion kept stable for 48 h and further increasing GA concentration to 5% led to improved stability for a week.

Although GA has been widely used in emulsions, since it's surface activity is very limited, more force is required during emulsion preparation. To overcome this shortcoming, some proteins have been used in combination with GA as co-emulsifiers (Gharibzahedi et al., 2013). Generally, the main emulsification effect was contributed by proteins, whereas, the stability of emulsion was provided by polysaccharides such as GA, owing to their steric and thickening effect, and meanwhile the GA not adsorbed onto the interface of the droplets could also enhance the solution viscosity and thus prevent phase separation. In addition, GA, and some other polysaccharides were used as co-emulsifiers or co-stabilizers while their enhanced solution viscosity could provide steric stabilization effect while their negative charge could provide electrostatic repulsion. For instance, Wang et al., reported the stability of flaxseed protein and soybean protein stabilized emulsion was significantly improved by adding 2% GA (Wang et al., 2011). The absorbed GA on an oil-water interface could provide the steric repulsion and improve the viscosity of emulsion leading to a better stability against ion strength and temperature change. Several applications of GA or GA containing co-emulsifiers in the formulations of emulsions are listed in Table 2.1.

In recent years, Pickering emulsion, stabilized by colloidal particles adsorption to the oil-water interface, has attracted increasing interests due to their "surfactant-free" advantage and sustained delivery behavior. Dai et al., prepared the Pickering emulsion and the colloidal complex nanoparticles consisted of zein and GA, which formed core-shell structure via electrostatic interactions and hydrogen bonding (Dai et al., 2018), were adopted to stabilize the emulsion. The particle size of GA-zein complex was approximately 200 nm, and it protected Pickering emulsion against

TABLE 2.1 Application of GA or GA Containing Complex in the Formulation of Emulsions

Materials used	Functions	References
GA	Used as an emulsifier to improve stability	Hosseini et al., 2015
	Used as a suspending agent to prevent the aggregation of carbon nanotubes and improve the stability	Alpatova et al., 2010; Singh et al., 2012
	Used as a stabilizer to stabilize gold or silver nanoparticles and improve the stability	Kattumuri et al., 2007; Martin et al., 2017
GA–whey protein isolate	Used as an emulsifier and improve the stability of Vitamin E emulsions against ion strength and temperature	Ozturk et al., 2015
GA–α-tocopherol/tertiary butyl hydroquinone/ascorbyl palmitate	Used as an emulsifier and protectβ-carotene from oxidation or degradation	Bouyer et al., 2011a
GA–ovalbumin	Used as an emulsifier and improve the stability of thyme oil	Niu et al., 2016
GA–xanthan	Used as a synergistic emulsifier and improve viscosity and stability of the emulsions	Desplanques et al., 2012
Lysozyme/acrylic-modified GA	Used as a thickening agent	Hashemi et al., 2018; Nickzare et al., 2009

coalescence and Ostwald ripening for 30 days under room temperature (Figure 2.1). Furthermore, the $O_1/W/O_2$ Pickering double emulsions were also prepared and stabilized by the combination of whey protein and GA and yield 76 days of storage stability under room temperature (Estrada-Fernandez et al., 2018).

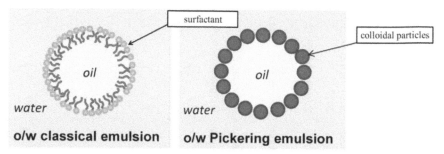

FIGURE 2.1 (See color insert.) Structure diagram of Pickering emulsion and conventional emulsion. Modified with permission from Chevalier and Bolzinger, 2013. © Elsevier.

2.3.2 GA AS AN EXCIPIENT IN TABLETS

GA has extensive application in tablets. It has been widely used in oral tablets as binders, and the tablets could disintegrate with short disintegration time and fast dissolution rate owing to the low viscosity, fast erosion and easily dissolving in water (Gowthamarajan et al., 2011; Hirata, 1970; Ogunjimi and Alebiowu, 2014). GA has also been employed as a release modifier or matrices for tablets preparation. Streubel et al., prepared verapamil and diltiazem extend-release floating tablets using HPMC alone and combination with GA as matrix-forming polymers to control the release rate of the model drugs (Streubel et al., 2003). Compared to the HPMC formulation, the tablets consisted of HPMC, and GA exhibited higher porosity and could maintain floating for at least 8 h in 0.1M HCl contributed to the fast dissolve of GA. However, this fast dissolving property also led to complete drug release within 2 hours, which was a disadvantage in controlled-release tablets. Furthermore, to enhance the controlled release properties of GA, acrylamide was grafted onto GA at 5:1 ratio to form a controlled release tablets matrix and the release of loaded diltiazem hydrochloride and nifedipine lasted up to 6 and 14 h, respectively (Toti et al., 2004).

2.3.3 GA AS THE WALL FORMING MATERIAL OF MICROCAPSULES

Microcapsules are microparticles with a core-shell structure; it is mainly composed of active ingredients and wall forming materials. The active ingredients may exist in the core in the form of solid or liquid (Campos et al., 2013). Both synthetic and natural polymers could be employed as wall forming materials. Microcapsules could provide many advantages such as targeting properties, controlled release of incorporated drugs, long shelf-life time and prevent side effects, improve the stability to the environment (Butstraen and Salaün, 2014; Huang et al., 2007).

Due to the non-toxicity, film-forming ability, high biocompatibility, biodegradability, and negative charge properties, GA has been widely used as the wall forming the material of microcapsules either alone or in combination with other polymers (Butstraen and Salaün, 2014). For instance, GA has been used to prepare microcapsules alone by spray drying to protect the natural beetroot juice against oxidation (Pitalua et al., 2010), and it was also employed as the shell material of riboflavin loaded microcapsule, with the photodegradation of riboflavin decreased by approximately 26% (Boiero et al., 2014).

GA can not only be used as a single carrier of microcapsules, due to its negative charge, but the combination of GA with some positively charged proteins and polysaccharides as the complex wall forming materials has also been extensively reported. The microcapsules consisted of GA and other materials as complex wall forming materials are usually prepared by the coacervation method. Compared to the microcapsules prepared by GA solely, the microcapsules consist of GA and other materials usually presented a more tightened structure, leading to higher drug loading efficiency and improved system stability (Baracat et al., 2012). For example, Shahgholian and Rajabzadeh fabricated and characterized curcumin-loaded bovine serum albumin (BSA) /GA microcapsules and BSA/GA weight ratio 2:1, pH 3.7 was selected. The drug encapsulation efficiency was 92% and significantly improved stability of curcumin due to the encapsulation (Shahgholian and Rajabzadeh, 2016). Similarly, GA could form microcapsules in combination with whey protein isolate via coacervation method (Eratte et al., 2017) and the probiotic bacteria *L. casei* and omega-3rich tuna oil was co-microencapsulated. The co-microencapsulation increased the survival of *L. casei* significantly as well as the surface hydrophobicity; therefore, adherence to the intestinal wall. Several other applications of GA based microcapsules are summarized in Table 2.2.

TABLE 2.2 Various Applications of GA Based Microcapsules Prepared by Coacervation Method

GA based wall forming materials	Specific functions	References
GA-gelation B	To achieve sustained release of model drugs, stronger stabilizing capability and switchable mechanical properties for self-healing, improve the drug loading efficiency	Chang et al., 2006; Kanellopoulos et al., 2017; Quan et al., 2013
GA-gelation A	To achieve controlled release properties	Bezerra et al., 2016
GA-maltodextrin	To improve the stability and protect model drugs against oxidation, light, moisture, and temperature	Tolun et al., 2016
GA-chitosan	To achieve mucoadhesive properties	Sakloetsakun et al., 2016
GA-almond gum	To improve the storage stability	Mahfoudhi and Hamdi, 2014
GA-hydrophobic derivatives of guar gum hydrolyzate	To improve the retention time of mint oil	Sarkar et al., 2013
GA-Whey Protein Isolate	To improve the stability of Vitamin C, omega-3 fatty acids and achieve controlled release of β-Carotene	Al-Ismail et al., 2016; Eratte et al., 2015; Jain et al., 2015

2.3.4   GA AS THE CARRIER OF MICROSPHERES

Recently, microspheres are becoming a promising vehicle for drug delivery contributed to the protective effect for the encapsulated drugs which are sensitive to the environment including moisture, oxygen, light, pH, and temperature (Fang and Bhandari, 2011). Different to microcapsules, which consist of a core and a shell, microspheres are composed of a homogeneous mixture of the active ingredient and matrix materials (Campos et al., 2013; Jiang et al., 2005). Interestingly, GA, due to its high water solubility, good emulsifying capacity, and low solution viscosity even at high concentration has become a potential carrier for microsphere preparation (Singh et al., 2007; Turchiuli et al., 2005).

GA can be used as a single matrix to prepare microspheres. For example, anthocyanin loaded GA microspheres were prepared (Carvalho et al., 2016), which could protect anthocyanin from oxidation, with over 88% of anthocyanin retained during preparation and storage. Similarly, carvacrol was encapsulated into GA microspheres with 93% encapsulation efficiency and improved stability (Da Costa et al., 2013).

Moreover, the combination of GA with other materials as the matrix material of microspheres to encapsulate active ingredients was more frequently reported. Compared with the single GA based microspheres, a controlled release behavior could be achieved by using complex materials since the viscosity of GA alone is very low. For example, Outuki et al., fabricated microspheres using a mixture of GA and xanthan gum by a spray drying method (Outuki et al., 2016). Rutinand hyperoside were incorporated in the microspheres, both protective effect against thermo and oxygen and sustained drug release behavior were achieved. However, it was also reported that the presence of GA could lead to a fast release behavior owing to its low viscosity. For instance, GA was sprayed with maltodextrin to prepare guaraná seeds to extract microsphere (Klein et al., 2015). The microspheres had an encapsulation efficiency of 55–60% with a particle size around 4.5 μm, the fabricated microsphere also improve the antioxidant capacity and protect the guaraná seeds extract from high temperature during the preparation process. Moreover, the model drug was released completely within 60 min.

2.3.5   GA AS THE CARRIER OF NANOPARTICLES

Nanoparticles, which is defined as particles rangingfrom1 to 1000 nm, are now becoming a promising drug delivery carrier with outstanding advantages of controlled drug delivery, improved solubility, prolonged blood circulation half-life, decreased side effect and improved biological efficiencies (Cheng et al., 2014; Ensign et al., 2012). Recently, polysaccharide-based nanoparticles have shown huge potential in delivering pharmaceutical active ingredients especially biomacromolecules including proteins, peptides, DNA, and biosensor (Jayakumar et al., 2010) since the nanoparticles can be prepared under gentle conditions. Notably, GA, as a biocompatible polymer with a negative charge, is frequently used in combination with other positively charged polymers to prepare nanoparticles (Liu et al., 2008).

For instance, GA, and positively charged CS were used in combination to prepare nanoparticles loading insulin, and an average diameter of 177–192 nm was obtained (Avadi et al., 2010). In the presence of GA, a faster insulin release rate was observed at higher pH value. This may be explained by that GA has a pH value of 4.5–5.0 in 5% W/V aqueous solution, and therefore the polymer chains may swell in the medium with pH higher than 6.5 and result in increased porosity in nanoparticle structure, leading to increased insulin release. Similar results were also reported by Avadi et al., (2011). Also, Tsai et al. fabricated various insulin-loaded CS/pectin/GA nanoparticles by complex coacervation method (Tsai et al., 2014). The prepared nanoparticles formed a globe-like network structure via molecular entanglement with the inner core consisted of GA/CS, and the outer part was mainly composed of CS and pectin, which could protect the nanoparticle from degradation in the GI tract; meanwhile, a controlled release behavior of loaded insulin was obtained. Notably, compared to the formulation without GA, the nanoparticles in the presence of GA exhibited a steady and fairly complete release of insulin in 6 h.

2.3.6   GA AS A COATING MATERIAL OF NANOPARTICLES

Recently, GA was also employed as a coating material to prepare gallic acid coated magnetic nanoparticles (GA-MNP) and applied in magnetic targeting and tumor imaging simultaneously via intravenous injection (Zhang et al., 2009). The coated GA on the surface of nanoparticles had two major advantages including bringing the functional conjugation groups for targeting ligands and provides steric hindrance to resist RES clearance so as to improve the *in vivo* stability and prolong circulation half-life. For example, Zhang et al., prepared GA coated magnetic nanoparticles with a mean hydrodynamic diameter of 118 ± 12 nm and presented a core-shell structure. The cell uptake results indicated that the GA-MNP was internalized into 9Lglioma cells and stayed in cytosol or endosome due to the specific particle size and the good biocompatibility of GA on the surface of nanoparticles. Similarly, Banerjee and Chen also modified the magnetic nanoparticles with GA and grafted with 2-hydroxypropyl-cyclodextrins as a new nanocarrier to deliver hydrophobic drugs (Banerjee and Chen, 2010).

Furthermore, it was also reported that using GA as a coating material of nanoparticles could improve its stability against aggregation so that the nanoparticles could have enough circulation time in physiological conditions. For instance, de Barros et al., prepared GA coated gold nanoparticles to prevent the aggregation and yield remarkable stability in physiological pH and even when the zeta potential was near zero, the gold nanoparticles exhibited surprisingly high stability (De Barros et al., 2016). In another study, epirubicin was loaded in GA coated gold nanoparticles functionalized with folic acid for tumor-targeted delivery (Devi et al., 2015). The cytotoxic effect of the nanoparticles on lung adenocarcinoma (A549) cells was enhanced, probably contributed to the increased uptake of epirubicin by cancer cells owing to cancer targeting properties and significantly improved stability of gold nanoparticle by GA coating.

2.3.7 GA AS A HYDROGEL INGREDIENT

Hydrogel has attracted increasing attention nowadays owing to its wide application in pharmaceuticals, biomedicine, catalysis, optics, etc. (Juby et al., 2012). The hydrogel could absorb a large amount of water and swell with its three-dimensional network structure maintained (Wu et al., 2010). Generally, the commonly used polymers for hydrogel preparation are natural or synthetic materials, and the polymers can be used alone or in combination. The most widely used natural materials are HA, alginate, CS, pectin, and collagen while the commonly used synthetic polymers are polyesters such as PEG-PLA-PEG, PEG-PCL-PEG, or PLA-PEG-PLA (Hamidi et al., 2008). Recently, GA has attracted increasingly interest as a hydrogel-forming material and used in tissue-engineering, encapsulating cells and bioactive molecules or wound dressings (Bonifacio et al., 2017; Fan et al., 2017) due to its excellent physicochemical properties such as non-toxicity, hemostatic, non-hemolytic, antibacterial, antioxidant, and anti-inflammatory (Singh and Dhiman, 2016). However, GA alone is not possible to form hydrogel since it could not gel by ion-interaction or pH change, and the low viscosity of GA solution makes it hard to form hydrogels. So, GA is usually used in combination with other polysaccharides or modified by other polymers to form hydrogels (Shankar et al., 2017).

It was reported that the release profiles and swelling degree of hydrogels could be improved in the presence of GA. For instance, Juby et al., prepared silver nanoparticles loaded PVA-GA hydrogel by a gamma radiation method (Juby et al., 2012). The nanoparticles with various particle sizes ranging from 10 to 40 nm were successfully loaded in the hydrogel. The equilibrium degree of swelling and release rate of silver increased significantly with the increase of GA amount, contributed to the enhanced hydrogen bonding with water and increased the hydrophilicity of the matrix. In another study, in order to improve the wound healing potential, the moxifloxacin loaded hydrogels consisted of GA, polyvinyl pyrrolidone, and carbopol were fabricated (Singh et al., 2017). The prepared hydrogels possess network structure and can absorb simulated wound fluid. Compared to the hydrogels without GA, the drug release was significantly increased in the simulated wound fluid, and the release behavior could be fitted into the non-Fickian diffusion mechanism and Higuchi model without any burst effect. Compared to the hydrogels in the presence of GA, the inflammation of tissues treated with non-GA hydrogels lasted for 12 days. In contrast, the GA containing hydrogel treated wound sections showed almost negligible inflammation on day 12, indicating that adding GA in the hydrogel could accelerate wound healing effectively.

On the other hand, GA was also used to prepare hydrogels by cross-linking with other polymers to achieve sustained release behavior. For instance, Li et al., prepared a bioinspired hydrogel to consist of GA and alginate and controlled release behavior was obtained (Li et al., 2017). The hydrogel was cross-linked by Ca^{2+} between GA and alginate to delivery an important cell membrane repair protein, mitsugumin 53 (MG53). The prepared hydrogel showed enhanced adhesion characteristics and sustained drug delivery properties in chronic wound healing. Some other applications of GA based hydrogels are summarized in Table 2.3.

TABLE 2.3 Other Applications of GA Based Hydrogels

Hydrogels	Usage	Reference
GA-modified alginate	To prevent polyphenol from degradation	Tsai et al., 2017
GA-acrylamide	To achieve controlled release of potassium and phosphate	Zonatto et al., 2017
GA-carbopol-polyvinylimidazole	To achieve the synergic effects of the mucoadhesive, antimicrobial, and antioxidant	Singh and Dhiman, 2016
GA, N,'N'-dimethylacrylamide, and methacrylic acid	pH-responsive and sustained release	Reis et al., 2016
GA-whey protein	To alter the mechanical properties	Valim et al., 2009
GA-acrylic	Fast swelling with good mechanical stability	Favaro et al., 2008; Zohuriaan-Mehr et al., 2006

2.3.8 GA AS AN ADDITIVE IN FILMS

Generally, film-based drug delivery systems attracted concern extensively due to their better availability and patient compliance. GA, with its excellent properties such as improved biodegradability and anti-inflammation, is attracting more and more attention in the formulation of film. But due to its extremely low solution viscosity, low thermal stability and poor mechanical strength, application of GA alone in film formulation were very limited. Thus, it was usually used in combination with other polymers to achieve more desirable properties, including a more smoother morphology, higher elongation to break and better mechanical strength (Tejada et al., 2017). For example, GA was used to prepare films in combination with poly (vinyl alcohol) (PVA), or polyvinyl pyrrolidone iodine (PVP-I) to develop iodine-release systems (Ahmad et al., 2008). By adding GA, the mechanical strength and water resistance of the PVA-GA films were significantly improved, and the similar trend was also observed in the PVP-I-GA-PVA system.

2.3.9 MODIFIED GA CONJUGATES IN DRUG DELIVERY

Currently, polysaccharide conjugate pro-drugs are proposed with its advantages of enhancement of drug solubility, stability, and prolonged circulation lifetime or altered release behavior. Among the polysaccharides, GA, with its high water solubility, biodegradability, biocompatibility, and ease of drug conjugation became an attractive candidate for pro-drug preparation (Ehren-freund-Kleinman et al., 2002; Feng et al., 2017). For instance, Stefanovic et al., developed nystatin-GA conjugates, and the conjugates significantly improved the stability of nystatin during the storage condition (Stefanovic et al., 2013). Compared to the free nystatin, the nystatin-GA conjugates also decreased the hemolytic activity but retained the antifungal activities. Mean-while, ampicillin-conjugated GA microspheres were fabricated by Nishi et al., (2007a). The prepared microspheres exhibited slow and sustained release behavior in phosphate buffer with only 10–25% drug released in 10 days depending on the drug payload and degree of GA conjugation but a fast release in simulated gastric fluid (SGF) due to rapid hydrolysis of the Schiff's linkage between ampicillin and GA. The group of Nishi et al., also developed an injectable self-gelling primaquine-GA conjugate and prima-quine-conjugated GA microspheres for controlled release of primaquine (Nishi and Jayakrishnan, 2004, 2007). Furthermore, Sarika et al., prepared GA-curcumin conjugate micelles to delivery curcumin to hepatocarcinoma cells (HepG2) and breast carcinoma cells (MCF-7) (Sarika et al., 2015). The micelles presented spherical morphology with a particle size of 270 ± 5 nm, a 900-fold higher solubility than that of free curcumin, and the micelles exhibited higher cytotoxicity to HepG2 cells than MCF-7 cells due to the targeting effect of galactose contributed to GA conjugation.

2.4 OTHER APPLICATIONS

2.4.1 GA AS A SHELF-LIFE ENHANCER

GA was also widely used in the food industry as an edible coat to enhance the shelf-life time of vegetables and fruits, which could retain freshness and quality (Maqbool et al., 2011). It could be used alone or in combina-tion with other polysaccharides such as CS.

2.4.2 GA USED IN TUMOR IMAGING

Nowadays, GA was also reported to be used in tumor imaging (Axiak-Bechtel et al., 2014; Chanda et al., 2014; Zhang et al., 2009). By coating or conjugating GA onto the nanoparticles' surface, the *in-vivo* stability and circulation half-life could be altered, and GA could readily conjugate with fluorophore rhodamine B, which could provide imaging effect. The tumor imaging could be achieved by both the fluorescent material and the EPR effect of nanoparticles (Zhang et al., 2009).

2.5 CONCLUSIONS AND PERSPECTIVES

GA has attracted extensive attention to be used as a potential excipient candidate in the pharmaceutical field owing to the excellent properties, including biodegradability, biocompatibility, thickening, and stabilizing capacity, and negative charge. GA has been broadly employed in drug delivery systems, especially as emulsifier or stabilizer. It was also used as wall forming materials of microcapsules and matrix of microspheres to improve the stability of drugs as well as forming nanoparticles to achieve targeted drug delivery. GA based hydrogel drug delivery system is promising for wound dressing. Recently, GA-drug conjugates were used as a drug vehicle and exhibited an improved drug solubility, controlled release behavior, and tumor cell targeting effect.

Nevertheless, despite the wide usage of GA in formulation fields, the exploration of the drug delivery mechanism is still not sufficient, and the exact interaction between drugs and GA is still unknown. Due to the complexity of the source, seasons, and areas, the reproducibility of GA is also a problem. In conclusion, GA is a polysaccharide with promising application in drug delivery, and further investigation of GA properties and its function as a pharmaceutical excipient are absolutely essential in the near future.

KEYWORDS

- **drug delivery**
- **gum acacia**
- **gum Arabic**
- **natural polymer**

REFERENCES

Abd El-Mawla, A. M. A., & Osman, H. E. H., (2011). Effects of gum acacia aqueous extract on the histology of the intestine and enzymes of both the intestine and the pancreas of albino rats treated with Meloxicam. *Pharmacognosy Research, 3,* 114–121.

Agüero, L., Zaldivar-Silva, D., Peña, L., & Dias, M., (2017). Alginate microparticles as oral colon drug delivery device: A review. *Carbohydr Polym, 168,* 32–43.

Ahmad, S. I., Hasan, N., Abid, C. K. V. Z., & Mazumdar, N., (2008). Preparation and characterization of films based on crosslinked blends of gum acacia, polyvinylalcohol, and polyvinylpyrrolidone-iodine complex. *Journal of Applied Polymer Science, 109,* 775–781.

Akhtar, M., & Dickinson, E., (2007). Whey protein-maltodextrin conjugates as emulsifying agents: An alternative to gum Arabic. *Food Hydrocolloids, 21,* 607–616.

Ali, B. H., Ziada, A., & Blunden, G., (2009). Biological effects of gum Arabic: A review of some recent research. *Food and Chemical Toxicology, 47,* 1–8.

Al-Ismail, K., El-Dijani, L., Al-Khatib, H., & Saleh, M., (2016). Effect of microencapsulation of vitamin C with gum Arabic, whey protein Isolate and some blends on its stability. *J. Sci. Ind. Res., 75,* 176–180.

Alpatova, A. L., Shan, W. Q., Babica, P., Upham, B. L., Rogensues, A. R., Masten, S. J., Drown, E., Mohanty, A. K., Alocilja, E. C., & Tarabara, V. V., (2010). Single-walled carbon nanotubes dispersed in aqueous media via non-covalent functionalization: Effect of dispersant on the stability, cytotoxicity, and epigenetic toxicity of nanotube suspensions. *Water Res., 44,* 505–520.

Annison, G., Trimble, R. P., & Topping, D. L., (1995). Feeding Australian acacia gums and gum Arabic leads to non-starch polysaccharide accumulation in the cecum of rats. *The Journal of Nutrition, 125,* 283–292.

Assaf, S., Phillips, G., & Williams, P., (2005). Studies on acacia exudate gums. Part I: The molecular weight of gum exudate. *Food Hydrocolloids, 19,* 647–660.

Avadi, M. R., Sadeghi, A. M. M., Mohamadpour, D. N., Dinarvand, R., Atyabi, F., & Rafiee-Tehrani, M., (2011). *Ex vivo* evaluation of insulin nanoparticles using chitosan and Arabic gum. *ISRN Pharmaceutics,* 860109.

Avadi, M. R., Sadeghi, A. M. M., Mohammadpour, N., Abedin, S., Atyabi, F., Dinarvand, R., & Rafiee-Tehrani, M., (2010). Preparation and characterization of insulin nanoparticles using chitosan and Arabic gum with ionic gelation method. *Nanomedicine-Nanotechnology Biology and Medicine, 6,* 58–63.

Axiak-Bechtel, A. M., Upendran, A., Lattimer, J. C., Kelsey, J., Cutler, C. S., Selting, K. A., et al., (2014). Gum Arabic-coated radioactive gold nanoparticles cause no short-term local or systemic toxicity in the clinically relevant canine model of prostate cancer. *International Journal of Nanomedicine, 9,* 5001–5011.

Banerjee, S. S., & Chen, D. H., (2010). Grafting of 2-hydroxypropyl-î²-cyclodextrin on gum Arabic-modified iron oxide nanoparticles as a magnetic carrier for targeted delivery of hydrophobic anticancer drug. *International Journal of Applied Ceramic Technology, 7,* 111–118.

Baracat, M., Nakagawa, A., Casagrande, R., Georgetti, S., Verri, W., & De Freitas, O., (2012). Preparation and characterization of microcapsules based on biodegradable polymers: Pectin/casein complex for controlled drug release systems. *AAPS Pharm. Sci. Tech., 13*, 364–372.

Belbekhouche, S., Charaabi, S., Picton, L., Le Cerf, D., & Carbonnier, B., (2018). Glucose-sensitive polyelectrolyte microcapsules based on (alginate/chitosan) pair. *Carbohydr Polym, 184*, 144–153.

Beyer, M., Reichert, J., Heurich, E., Jandt, K. D., & Sigusch, B. W., (2010). Pectin, alginate and gum Arabic polymers reduce citric acid erosion effects on human enamel. *Dental Materials, 26*, 831–839.

Bezerra, F. M., Garcia, C. O., Garcia, C. C., Jose, L. M., & De Moraes, F. F., (2016). Controlled release of microencapsulated citronella essential oil on cotton and polyester matrices. *Cellulose, 23*, 1459–1470.

Boiero, M. L., Mandrioli, M., Vanden, B. N., Rodriguez-Estrada, M. T., García, N. A., Borsarelli, C. D., & Montenegro, M. A., (2014). Gum Arabic microcapsules as protectors of the photoinduced degradation of riboflavin in whole milk. *Journal of Dairy Science, 97*, 5328–5336.

Bonifacio, M. A., Gentile, P., Ferreira, A. M., Cometa, S., & De Giglio, E., (2017). Insight into halloysite nanotubes-loaded gellan gum hydrogels for soft tissue engineering applications. *Carbohydrate Polymers, 163*, 280–291.

Bouyer, E., Mekhloufi, G., Le Potier, I., De Kerdaniel, T. D. F., Grossiord, J. L., Rosilio, V., & Agnely, F., (2011a). Stabilization mechanism of oil-in-water emulsions by beta-lactoglobulin and gum Arabic. *Journal of Colloid and Interface Science, 354*, 467–477.

Bouyer, E., Mekhloufi, G., Le Potier, I., De Kerdaniel, T. F., Grossiord, J. L., Rosilio, V., & Agnely, F., (2011b). Stabilization mechanism of oil-in-water emulsions by beta-lactoglobulin and gum Arabic. *J Colloid Interface Sci., 354*, 467–477.

Butstraen, C., & Salaün, F., (2014). Preparation of microcapsules by complex coacervation of gum Arabic and chitosan. *Carbohydrate Polymers, 99*, 608–616.

Campos, E., Branquinho, J., Carreira, A. S., Carvalho, A., Coimbra, P., Ferreira, P., & Gil, M. H., (2013). Designing polymeric microparticles for biomedical and industrial applications. *European Polymer Journal, 49*, 2005–2021.

Carvalho, A. G. D., Machado, M. T. D., Da Silva, V. M., Sartoratto, A., Rodrigues, R. A. F., & Hubinger, M. D., (2016). Physical properties and morphology of spray dried microparticles containing anthocyanins of jussara (Euterpe edulis Martius) extract. *Powder Technol., 294*, 421–428.

Chanda, N., Upendran, A., Boote, E. J., Zambre, A., Axiak, S., Selting, K., Katti, K. V., Leevy, W. M., Afrasiabi, Z., Vimal, J., Singh, J., Lattimer, J. C., & Kannan, R., (2014). Gold nanoparticle-based x-ray contrast agent for tumor imaging in mice and dog: A potential nano-platform for computer tomography theranostics. *Journal of Biomedical Nanotechnology, 10*, 383–392.

Chang, C. P., Leung, T. K., Lin, S. M., & Hsu, C. C., (2006). Release properties on gelatin-gum Arabic microcapsules containing camphor oil with added polystyrene. *Colloids and Surfaces B-Biointerfaces, 50*, 136–140.

Chaturvedi, S., & Jones, C., (2007). *Protein Restriction for Children with Chronic Renal Failure.* Cochrane database of systematic reviews, *17*(4), CD006863.

Cheng, Y., Morshed, R., Auffinger, B., Tobias, A., & Lesniak, M., (2014). Multifunctional nanoparticles for brain tumor imaging and therapy. *Adv. Drug Deliv. Rev., 66*, 42–57.

Chevalier, Y., & Bolzinger, M. A., (2013). Emulsions stabilized with solid nanoparticles: Pickering emulsions. *Colloids and Surfaces A: Physicochemical and Engineering Aspects, 439*, 23–34.

Coviello, T., Matricardi, P., Marianecci, C., & Alhaique, F., (2007). Polysaccharide hydrogels for modified release formulations. *Journal of Controlled Release, 119*, 5–24.

Da Costa, J. M. G., Borges, S. V., Hijo, A., Silva, E. K., Marques, G. R., Cirillo, M. A., & De Azevedo, V. M., (2013). Matrix structure selection in the microparticles of essential oil oregano produced by spray dryer. *Journal of Microencapsulation, 30*, 717–727.

Dai, L., Sun, C., Wei, Y., Mao, L., & Gao, Y., (2018). Characterization of Pickering emulsion gels stabilized by zein/gum Arabic complex colloidal nanoparticles. *Food Hydrocolloids, 74*, 239–248.

De Barros, H. R., Cardoso, M. B., De Oliveira, C. C., Franco, C. R. C., Belan, D. D., Vidotti, M., & Riegel-Vidotti, I. C., (2016). Stability of gum Arabic-gold nanoparticles in physiological simulated pHs and their selective effect on cell lines. *Rsc. Advances, 6*, 9411–9420.

Desplanques, S., Renou, F., Grisel, M., & Malhiac, C., (2012). Impact of chemical composition of xanthan and acacia gums on the emulsification and stability of oil-in-water emulsions. *Food Hydrocolloids, 27*, 401–410.

Devi, P. R., Kumar, C. S., Selvamani, P., Subramanian, N., & Ruckmani, K., (2015). Synthesis and characterization of Arabic gum capped gold nanoparticles for tumor-targeted drug delivery. *Materials Letters, 139*, 241–244.

Dror, Y., Cohen, Y., & Yerushalmi-Rozen, R., (2006). Structure of gum Arabic in aqueous solution. *Journal of Polymer Science Part B: Polymer Physics, 44*, 3265–3271.

Eastwood, M. A., (1992). The physiological effect of dietary fiber: An update. *Annual Review of Nutrition, 12*, 19–35.

Ehrenfreund-Kleinman, T., Azzam, T., Falk, R., Polacheck, I., Golenser, J., & Domb, A. J., (2002). Synthesis and characterization of novel water-soluble amphotericin B-arabinogalactan conjugates. *Biomaterials, 23*, 1327–1335.

Ensign, L. M., Cone, R., & Hanes, J., (2012). Oral drug delivery with polymeric nanoparticles: The gastrointestinal mucus barriers. *Advanced Drug Delivery Reviews, 64*, 557–570.

Eratte, D., Dowling, K., Barrow, C. J., & Adhikari, B. P., (2017). In-vitro digestion of probiotic bacteria and omega-3 oil co-microencapsulated in whey protein isolate-gum Arabic complex coacervates. *Food Chemistry, 227*, 129–136.

Eratte, D., McKnight, S., Gengenbach, T. R., Dowling, K., Barrow, C. J., & Adhikari, B. P., (2015). Co-encapsulation and characterization of omega-3 fatty acids and probiotic bacteria in whey protein isolate-gum Arabic complex coacervates. *J. Funct. Food, 19*, 882–892.

Estrada-Fernandez, A. G., Roman-Guerrero, A., Jimenez-Alvarado, R., Lobato-Calleros, C., Alvarez-Ramirez, J., & Vernon-Carter, E. J., (2018). Stabilization of oil-in-water-in-oil (O-1/W/O-2) Pickering double emulsions by soluble and insoluble whey protein concentrate-gum Arabic complexes used as inner and outer interfaces. *Journal of Food Engineering, 221*, 35–44.

Fan, M., Ma, Y., Tan, H., Jia, Y., Zou, S., Guo, S., Zhao, M., Huang, H., Ling, Z., Chen, Y., & Hu, X., (2017). Covalent and injectable Chitosan-chondroitin sulfate hydrogels embedded with chitosan microspheres for drug delivery and tissue engineering. *Mater. Sci. Eng. C-Mater. Biol. Appl., 71*, 67–74.

Fang, Z., & Bhandari, B., (2011). Effect of spray drying and storage on the stability of bayberry polyphenols. *Food Chemistry, 129*, 1139–1147.

Favaro, S. L., De Oliveira, F., Reis, A. V., Guilherme, M. R., Muniz, E. C., & Tambourgi, E. B., (2008). Superabsorbent hydrogel composed of covalently cross-linked gum Arabic with fast swelling dynamics. *Journal of Applied Polymer Science, 107*, 1500–1506.

Feng, X. R., Li, D., Han, J. D., Zhuang, X. L., & Ding, J. X., (2017). Schiff base bond-linked polysaccharide-doxorubicin conjugate for upregulated cancer therapy. *Mater. Sci. Eng. C-Mater. Biol. Appl., 76*, 1121–1128.

Gharibzahedi, S. M. T., Razavi, S. H., & Mousavi, S. M., (2013). Ultrasound-assisted formation of the canthaxanthin emulsions stabilized by Arabic and xanthan gums. *Carbohydrate Polymers, 96*, 21–30.

Gowthamarajan, K., Kumar, G. K. P., Gaikwad, N. B., & Suresh, B., (2011). Preliminary study of Anacardium occidentale gum as binder in formulation of paracetamol tablets. *Carbohydrate Polymers, 83*, 506–511.

Hamidi, M., Azadi, A., & Rafiei, P., (2008). Hydrogel nanoparticles in drug delivery. *Advanced Drug Delivery Reviews, 60*, 1638–1649.

Hashemi, M. M., Aminlari, M., Forouzan, M. M., Moghimi, E., Tavana, M., Shekarforoush, S., & Mohammadifar, M. A., (2018). Production and application of lysozyme-gum Arabic conjugate in mayonnaise as a natural preservative and emulsifier. *Pol. J. Food Nutr. Sci., 68*, 33–43.

He, H., Zhang, M., Liu, L., Zhang, S., Liu, J., & Zhang, W., (2015). Suppression of remodeling behaviors with arachidonic acid modification for enhanced *in vivo* antiatherogenic efficacies of lovastatin-loaded discoidal recombinant high-density lipoprotein. *Pharmaceutical Research, 32*, 3415–3431.

Hirata, G., (1970). Interaction of drugs with polymers. VII. On the mechanism of tablet disintegration by crosslinked gum Arabic. Yakugaku Zasshi: *Journal of the Pharmaceutical Society of Japan, 90*, 661–664.

Hosseini, A., Jafari, S. M., Mirzaei, H., Asghari, A., & Akhavan, S., (2015). Application of image processing to assess emulsion stability and emulsification properties of Arabic gum. *Carbohydrate Polymers, 126*, 1–8.

Huang, J., Deng, Y., Ren, J., Chen, G., Wang, G., Wang, F., & Wu, X., (2018). Novel in situ forming hydrogel based on xanthan and chitosan re-gelifying in liquids for local drug delivery. *Carbohydr Polym, 186*, 54–63.

Huang, Y. I., Cheng, Y. H., Yu, C. C., Tsai, T. R., & Cham, T. M., (2007). Microencapsulation of extract containing shikonin using gelatin-acacia coacervation method: A formaldehyde-free approach. *Colloids and Surfaces B-Biointerfaces, 58*, 290–297.

Jafari, S. M., Assadpoor, E., He, Y., & Bhandari, B., (2008). Re-coalescence of emulsion droplets during high-energy emulsification. *Food Hydrocolloids, 22*, 1191–1202.

Jain, A., Thakur, D., Ghoshal, G., Katare, O. P., & Shivhare, U. S., (2015). Microencapsulation by complex coacervation using whey protein isolates and gum acacia: An approach to preserve the functionality and controlled release of beta-carotene. *Food and Bioprocess Technology, 8*, 1635–1644.

Jensen, C. D., Spiller, G. A., Gates, J. E., Miller, A. F., & Whittam, J. H., (1993). The effect of acacia gum and a water-soluble dietary fiber mixture on blood lipids in humans. *Journal of the American College of Nutrition, 12*, 147–154.

Ji, C. D., Khademhosseini, A., & Dehghani, F., (2011). Enhancing cell penetration and proliferation in chitosan hydrogels for tissue engineering applications. *Biomaterials, 32,* 9719–9729.

Jiang, W. L., Gupta, R. K., Deshpande, M. C., & Schwendeman, S. P., (2005). Biodegradable poly(lactic-co-glycolic acid) microparticles for injectable delivery of vaccine antigens. *Advanced Drug Delivery Reviews, 57,* 391–410.

Juby, K. A., Dwivedi, C., Kumar, M., Kota, S., Misra, H. S., & Bajaj, P. N., (2012). Silver nanoparticle-loaded PVA/gum acacia hydrogel: Synthesis, characterization and antibacterial study. *Carbohydrate Polymers, 89,* 906–913.

Kanellopoulos, A., Giannaros, P., Palmer, D., Kerr, A., & Al-Tabbaa, A., (2017). Polymeric microcapsules with switchable mechanical properties for self-healing concrete: Synthesis, characterization and proof of concept. *Smart Materials and Structures, 26.*

Kattumuri, V., Katti, K., Bhaskaran, S., Boote, E. J., Casteel, S. W., Fent, G. M., Robertson, D. J., Chandrasekhar, M., Kannan, R., & Katti, K. V., (2007). Gum Arabic as a phytochemical construct for the stabilization of gold nanoparticles: *In vivo* pharmacokinetics and X-ray-contrast-imaging studies. *Small, 3,* 333–341.

Klein, T., Longhini, R., Bruschi, M. L., & De Mello, J. C. P., (2015). Microparticles containing guaraná extract obtained by spray-drying technique: Development and characterization. *Revista Brasileira de Farmacognosia, 25,* 292–300.

Krishnan, S., Bhosale, R., & Singhal, R. S., (2005). Microencapsulation of cardamom oleoresin: Evaluation of blends of gum arabic, maltodextrin and a modified starch as wall materials. *Carbohydrate Polymers, 61,* 95–102.

Lasareva, E., Chernuchina, A., & Gabrielyan, G., (2018). Preparation, surface activity and colloidal properties of the ionic complex of chitosan with hexadecyl-oligo-oxyethylene hemisuccinate. *Carbohydr Polym, 183,* 123–130.

Leclere, C. J., Champ, M., Boillot, J., Guille, G., Lecannu, G., Molis, C., Bornet, F., Krempf, M., Delort-Laval, J., & Galmiche, J. P., (1994). Role of viscous guar gums in lowering the glycemic response after a solid meal. *The American Journal of Clinical Nutrition, 59,* 914–921.

Leroux, J., Langendorff, V., Schick, G., Vaishnav, V., & Mazoyer, J., (2003). Emulsion stabilizing properties of pectin. *Food Hydrocolloids, 17,* 455–462.

Li, L., Ni, R., Shao, Y., & Mao, S., (2014). Carrageenan and its applications in drug delivery. *Carbohydr. Polym., 103,* 1–11.

Li, M., Li, H., Li, X., Zhu, H., Xu, Z., Liu, L., Ma, J., & Zhang, M., (2017). A bioinspired alginate-gum Arabic hydrogel with micro-/nanoscale structures for controlled drug release in chronic wound healing. *ACS Applied Materials & Interfaces, 9,* 22160–22175.

Liu, Y., Hou, Z., Yang, J., & Gao, Y., (2015). Effects of antioxidants on the stability of β-carotene in O/W emulsions stabilized by Gum Arabic. *J. Food Sci. Technol., 52,* 3300–3311.

Liu, Z. H., Jiao, Y. P., Wang, Y. F., Zhou, C. R., & Zhang, Z. Y., (2008). Polysaccharides-based nanoparticles as drug delivery systems. *Advanced Drug Delivery Reviews, 60,* 1650–1662.

Mahfoudhi, N., & Hamdi, S., (2014). Kinetic degradation and storage stability of beta-carotene encapsulated by spray drying using almond gum and gum Arabic as wall materials. *J. Polym. Eng., 34,* 683–693.

Mao, S., Sun, W., & Kissel, T., (2010). Chitosan-based formulations for delivery of DNA and siRNA. *Adv. Drug Deliv. Rev., 62,* 12–27.

Maqbool, M., Ali, A., Alderson, P. G., Zahid, N., & Siddiqui, Y., (2011). Effect of a novel edible composite coating based on gum Arabic and chitosan on biochemical and physiological responses of banana fruits during cold storage. *Journal of Agricultural and Food Chemistry, 59*, 5474–5482.

Martin, J. D., Telgmann, L., & Metcalfe, C. D., (2017). A method for preparing silver nanoparticle suspensions in bulk for ecotoxicity testing and ecological risk assessment. *Bull. Environ. Contam. Toxicol., 98*, 589–594.

Matsumoto, N., Riley, S., Fraser, D., Al-Assaf, S., Ishimura, E., Wolever, T., Phillips, G. O., & Phillips, A. O., (2006). Butyrate modulates TGF-beta 1 generation and function: Potential renal benefit for Acacia(sen) SUPERGUM (TM) (gum Arabic)? *Kidney International, 69*, 257–265.

McClements, D. J., (2012). Advances in fabrication of emulsions with enhanced functionality using structural design principles. *Current Opinion in Colloid & Interface Science, 17*, 235–245.

Mirtič, J., Ilaš, J., & Kristl, J., (2018). Influence of different classes of crosslinkers on alginate polyelectrolyte nanoparticle formation, thermodynamics and characteristics. *Carbohydr. Polym., 181*, 93–102.

Mochida, S., Ogata, I., Hirata, K., Ohta, Y., Yamada, S., & Fujiwara, K., (1990). Provocation of massive hepatic necrosis by endotoxin after partial hepatectomy in rats. *Gastroenterology, 99*, 771–777.

Mochida, S., Ohno, A., Arai, M., Tamatani, T., Miyasaka, M., & Fujiwara, K., (1996). Role of adhesion molecules in the development of massive hepatic necrosis in rats. *Hepatology (Baltimore, Md.), 23*, 320–328.

Moundras, C., Behr, S. R., Demigne, C., Mazur, A., & Remesy, C., (1994). Fermentable polysaccharides that enhance fecal bile acid excretion lower plasma cholesterol and apolipoprotein E-rich HDL in rats. *The Journal of Nutrition, 124*, 2179–2188.

Nasir, O., Artunc, F., Saeed, A., Kambal, M. A., Kalbacher, H., Sandulache, D., Boini, K. M., Jahovic, N., & Lang, F., (2008). Effects of gum Arabic (Acacia senegal) on water and electrolyte balance in healthy mice. *Journal of Renal Nutrition, 18*, 230–238.

Nasir, O., Umbach, A. T., Rexhepaj, R., Ackermann, T. F., Bhandaru, M., Ebrahim, A., et al., (2012). Effects of gum Arabic (Acacia senegal) on renal function in diabetic mice. *Kidney & Blood Pressure Research, 35*, 365–372.

Nickzare, M., Zohuriaan-Mehr, M. J., Yousefi, A. A., & Ershad-Langroudi, A., (2009). Novel acrylic-modified acacia gum thickener: Preparation, characterization and rheological properties. *Starch – Stärke, 61*, 188–198.

Nishi, K. K., & Jayakrishnan, A., (2004). Preparation and in vitro evaluation of primaquine-conjugated gum Arabic microspheres. *Biomacromolecules, 5*, 1489–1495.

Nishi, K. K., & Jayakrishnan, A., (2007). Self-gelling primaquine-gum Arabic conjugate: An injectable controlled delivery system for primaquine. *Biomacromolecules, 8*, 84–90.

Nishi, K. K., Antony, M., & Jayakrishnan, A., (2007a). Synthesis and evaluation of ampicillin-conjugated gum Arabic microspheres for sustained release. *Journal of Pharmacy and Pharmacology, 59*, 485–493.

Nishi, K. K., Antony, M., Mohanan, P. V., Anilkumar, T. V., Loiseau, P. M., & Jayakrishnan, A., (2007b). Amphotericin B-gum Arabic conjugates: Synthesis, toxicity, bioavailability, and activities against Leishmania and fungi. *Pharmaceutical Research, 24*, 971–980.

Niu, F., Pan, W., Su, Y., & Yang, Y., (2016). Physical and antimicrobial properties of thyme oil emulsions stabilized by ovalbumin and gum Arabic. *Food Chemistry, 212*, 138–145.

Ogunjimi, A. T., & Alebiowu, G., (2014). Neem gum as a binder in a formulated paracetamol tablet with reference to acacia gum BP. *AAPS Pharm. Sci. Tech., 15*, 500–510.

Onishi, T., Umemura, S., Yanagawa, M., Matsumura, M., Sasaki, Y., Ogasawara, T., & Ooshima, T., (2008). Remineralization effects of gum Arabic on caries-like enamel lesions. *Archives of Oral Biology, 53*, 257–260.

Outuki, P. M., Belloto de Francisco, L. M., Hoscheid, J., Bonifacio, K. L., Barbosa, D. S., & Carvalho, C. M. L., (2016). Development of Arabic and xanthan gum microparticles loaded with an extract of Eschweilera nana Miers leaves with antioxidant capacity. *Colloids and Surfaces a-Physicochemical and Engineering Aspects, 499*, 103–112.

Ozturk, B., Argin, S., Ozilgen, M., & McClements, D. J., (2015). Formation and stabilization of nanoemulsion-based vitamin E delivery systems using natural biopolymers: Whey protein isolate and gum Arabic. *Food Chemistry, 188*, 256–263.

Pasquier, B., Armand, M., Castelain, C., Guillon, F., Borel, P., Lafont, H., & Lairon, D., (1996). Emulsification and lipolysis of triacylglycerols are altered by viscous soluble dietary fibers in acidic gastric medium *in vitro*. *The Biochemical Journal, 314*(Part 1), 269–275.

Patel, S., & Goyal, A., (2015). Applications of natural polymer gum Arabic: A review. *International Journal of Food Properties, 18*, 986–998.

Paulino, A. T., Guilherme, M. R., Reis, A. V., Campese, G. M., Muniz, E. C., & Nozaki, J., (2006). Removal of methylene blue dye from an aqueous media using superabsorbent hydrogel supported on modified polysaccharide. *Journal of Colloid and Interface Science, 301*, 55–62.

Phillips, G. O., (1998). Acacia gum (Gum Arabic): A nutritional fiber, metabolism and calorific value. *Food Additives and Contaminants, 15*, 251–264.

Pitalua, E., Jimenez, M., Vernon-Carter, E. J., & Beristain, C. I., (2010). Antioxidative activity of microcapsules with beetroot juice using gum Arabic as wall material. *Food and Bioproducts Processing, 88*, 253–258.

Pizzoni, D., Compagnone, D., Di Natale, C., D'Alessandro, N., & Pittia, P., (2015). Evaluation of aroma release of gummy candies added with strawberry flavors by gas-chromatography/mass-spectrometry and gas sensors arrays. *Journal of Food Engineering, 167*, 77–86.

Pradeep, A. R., Happy, D., & Garg, G., (2010). Short-term clinical effects of commercially available gel containing Acacia arabica: A randomized controlled clinical trial. *Aust. Dent. J., 55*, 65–69.

Quan, J., Kim, S. M., Pan, C. H., & Chung, D., (2013). Characterization of fucoxanthin-loaded microspheres composed of cetyl palmitate-based solid lipid core and fish gelatin-gum Arabic coacervate shell. *Food Research International, 50*, 31–37.

Reis, A. V., Moia, T. A., Sitta, D. L. A., Mauricio, M. R., Tenorio-Neto, E. T., Guilherme, M. R., Rubira, A. F., & Muniz, E. C., (2016). Sustained release of potassium diclofenac from a pH-responsive hydrogel based on gum Arabic conjugates into simulated intestinal fluid. *Journal of Applied Polymer Science, 133*.

Ross, A. H., Eastwood, M. A., Brydon, W. G., Anderson, J. R., & Anderson, D. M., (1983). A study of the effects of dietary gum Arabic in humans. *The American Journal of Clinical Nutrition, 37*, 368–375.

Sakloetsakun, D., Preechagoon, D., Bernkop-Schnurch, A., & Pongjanyakul, T., (2016). Chitosan-gum Arabic polyelectrolyte complex films: Physicochemical, mechanical and mucoadhesive properties. *Pharm. Dev. Technol., 21*, 590–599.

Sarika, P. R., James, N. R., Kumar, P. R., Raj, D. K., & Kumary, T. V., (2015). Gum arabic-curcumin conjugate micelles with enhanced loading for curcumin delivery to hepatocarcinoma cells. *Carbohydr Polym., 134*, 167–174.

Sarkar, S., Gupta, S., Variyar, P. S., Sharma, A., & Singhal, R. S., (2013). Hydrophobic derivatives of guar gum hydrolyzate and gum Arabic as matrices for microencapsulation of mint oil. *Carbohydrate Polymers, 95*, 177–182.

Shahgholian, N., & Rajabzadeh, G., (2016). Fabrication and characterization of curcumin-loaded albumin/gum Arabic coacervate. *Food Hydrocolloids, 59*, 17–25.

Shankar, K. G., Gostynska, N., Montesi, M., Panseri, S., Sprio, S., Kon, E., Marcacci, M., Tampieri, A., & Sandri, M., (2017). Investigation of different cross-linking approaches on 3D gelatin scaffolds for tissue engineering application: A comparative analysis. *International Journal of Biological Macromolecules, 95*, 1199–1209.

Singh, B. P., Nayak, S., Samal, S., Bhattacharjee, S., & Besra, L., (2012). Characterization and dispersion of multiwalled carbon nanotubes (MWCNTs) in aqueous suspensions: Surface chemistry aspects. *Journal of Dispersion Science and Technology, 33*, 1021–1029.

Singh, B., & Dhiman, A., (2016). Design of acacia gum-carbopol-cross-linked-polyvinylimidazole hydrogel wound dressings for antibiotic/anesthetic drug delivery. *Industrial & Engineering Chemistry Research, 55*, 9176–9188.

Singh, B., Sharma, S., & Dhiman, A., (2017). Acacia gum polysaccharide-based hydrogel wound dressings: Synthesis, characterization, drug delivery and biomedical properties. *Carbohydrate Polymers, 165*, 294–303.

Singh, M., Chakrapani, A., & O'Hagon, D., (2007). Nanoparticles and microparticles as vaccine-delivery systems. *Expert Rev. Vaccines, 6*, 797–808.

Stefanovic, J., Jakovljevic, D., Gojgic-Cvijovic, G., Lazic, M., & Vrvic, M., (2013). Synthesis, characterization, and antifungal activity of nystatingum Arabic conjugates. *Journal of Applied Polymer Science, 127*, 4736–4743.

Streubel, A., Siepmann, J., & Bodmeier, R., (2003). Floating matrix tablets based on low-density foam powder: Effects of formulation and processing parameters on drug release. *European Journal of Pharmaceutical Sciences, 18*, 37–45.

Tahmassebi, J. F., Duggal, M. S., Malik-Kotru, G., & Curzon, M. E. J., (2006). Soft drinks and dental health: A review of the current literature. *Journal of Dentistry, 34*, 2–11.

Tejada, G., Piccirilli, G. N., Sortino, M., Salomón, C. J., Lamas, M. C., & Leonardi, D., (2017). Formulation and *in-vitro* efficacy of antifungal mucoadhesive polymeric matrices for the delivery of miconazole nitrate. *Materials Science and Engineering: C., 79*, 140–150.

Tolun, A., Altintas, Z., & Artik, N., (2016). Microencapsulation of grape polyphenols using maltodextrin and gum Arabic as two alternative coating materials: Development and characterization. *J. Biotechnol., 239*, 23–33.

Toti, U. S., Soppimath, K. S., Mallikarjuna, N. N., & Aminabhavi, T. M., (2004). Acrylamide-grafted-acacia gum polymer matrix tablets as erosion-controlled drug delivery systems. *Journal of Applied Polymer Science, 93*, 2245–2253.

Tsai, F. H., Kitamura, Y., & Kokawa, M., (2017). Effect of gum Arabic-modified alginate on physicochemical properties, release kinetics, and storage stability of liquid-core hydrogel beads. *Carbohydrate Polymers, 174*, 1069–1077.

Tsai, R. Y., Chen, P. W., Kuo, T. Y., Lin, C. M., Wang, D. M., Hsien, T. Y., & Hsieh, H. J., (2014). Chitosan/pectin/gum Arabic polyelectrolyte complex: Process-dependent appearance, microstructure analysis and its application. *Carbohydrate Polymers, 101,* 752–759.

Turchiuli, C., Fuchs, M., Bohin, M., Cuvelier, M. E., Ordonnaud, C., Peyrat-Maillard, M. N., & Dumoulin, E., (2005). Oil encapsulation by spray drying and fluidized bed agglomeration. *Innovative Food Science & Emerging Technologies 6,* 29–35.

Ushida, K., Hatanaka, H., Inoue, R., Tsukahara, T., & Phillips, G. O., (2011). Effect of long term ingestion of gum Arabic on the adipose tissues of female mice. *Food Hydrocolloids, 25,* 1344–1349.

Valim, M. D., Cavallieri, A. L. F., & Cunha, R. L., (2009). Whey protein/Arabic gum gels formed by chemical or physical gelation process. *Food Biophys., 4,* 23–31.

Verbeken, D., Dierckx, S., & Dewettinck, K., (2003). Exudate gums: Occurrence, production, and applications. *Applied Microbiology and Biotechnology, 63,* 10–21.

Wadood, A., Wadood, N., & Shah, S. A., (1989). Effects of Acacia arabica and *Caralluma edulis* on blood glucose levels of normal and alloxan diabetic rabbits. *JPMA: The Journal of the Pakistan Medical Association, 39,* 208–212.

Wang, B., Wang, L. J., Li, D., Adhikari, B., & Shi, J., (2011). Effect of gum Arabic on stability of oil-in-water emulsion stabilized by flaxseed and soybean protein. *Carbohydrate Polymers, 86,* 343–351.

Weinbreck, F., De Vries, R., Schrooyen, P., & De Kruif, C. G., (2003). Complex conservation of whey proteins and gum Arabic. *Biomacromolecules, 4,* 293–303.

Winchester, J. F., & Salsberg, J. A., (2004). Sorbents in the treatment of renal failure. Minerva urologica e nefrologica. *The Italian Journal of Urology and Nephrology, 56,* 215–221.

Wu, Y., Xia, M., Fan, Q., & Zhu, M., (2010). Designable synthesis of nanocomposite hydrogels with excellent mechanical properties based on chemical cross-linked interactions. *Chemical Communications, 46,* 7790–7792.

Zhang, L., Yu, F., Cole, A. J., Chertok, B., David, A. E., Wang, J., & Yang, V. C., (2009). Gum Arabic-coated magnetic nanoparticles for potential application in simultaneous magnetic targeting and tumor imaging. *AAPS Journal, 11,* 693–699.

Zohuriaan-Mehr, M. J., Motazedi, Z., Kabiri, K., Ershad-Langroudi, A., & Allahdadi, I., (2006). Gum Arabic-acrylic superabsorbing hydrogel hybrids: Studies on swelling rate and environmental responsiveness. *Journal of Applied Polymer Science, 102,* 5667–5674.

Zonatto, F., Muniz, E. C., Tambourgi, E. B., & Paulino, A. T., (2017). Adsorption and controlled release of potassium, phosphate and ammonia from modified Arabic gum-based hydrogel. *International Journal of Biological Macromolecules, 105,* 363–369.

CHAPTER 3

Recent Advances in Pharmaceutical Applications of Natural Carbohydrate Polymer Gum Tragacanth

MADHUSMITA DHUPAL,[1] MUKESH KUMAR GUPTA,[2]
DIPTI RANJAN TRIPATHY,[3] MOHIT KUMAR,[4] DONG KEE YI,[5]
SITANSU SEKHAR NANDA,[5] and DEVASISH CHOWDHURY[6]

[1]*Department of Global Medical Science, Wonju College of Medicine, Yonsei University, Gangwon–220701, Republic of Korea*

[2]*Department of Biotechnology and Medical Engineering, National Institute of Technology, Rourkela, Odisha – 769008, India*

[3]*Department of Neurology, Gauhati Medical College and Hospital, Gauhati – 781032, Assam, India*

[4]*School of Advanced Materials Science and Engineering, Sungkyunkwan University, Suwon, South Korea*

[5]*Department of Chemistry, Myongji University, Yongin, South Korea*

[6]*Department of Material nanochemistry, Physical Science Division, Institute of Advanced Study in Science and Technology (IASST), Guwahati – 781035, Assam, India*

ABSTRACT

While naturally occurring biodegradable polymer materials are acquiring increasing attention in pharmaceutical industries owing to its burgeoning application with biocompatibility, non-toxicity, low cost, abundance, and high dispersibility, there have been circumscribed apprehension of the possible applications of Gum Tragacanth (GT) as an excipient for therapeutic enhancement of drug delivery system. Nevertheless, latterly, GT has been looked into as nanoparticles conjugated-cargos for mucosal, oral, nasal, topical therapeutic formulations. Fabricated GT with tunable functionality affords new exciting prospects extrapolating GT as a promising excipient compared to its counterparts. Macromolecular prodrug GT is a polysaccharide, swellable; bioadhesive to the target tissue, readily reabsorbed and its non-immunogenic physiochemical properties make it an excellent candidate for the nano-drug carrier. Therefore, this chapter concealments the elaborated overview of the most recent biomedical and pharmaceutical applications and challenges of natural carbohydrate polymer GT, in special reference to varied hydrogel-based tissue engineering, skin repair, nerve regeneration, bone repair, dental repair, nano-encapsulation, controlled site-specific drug delivery.

3.1   INTRODUCTION

The roles of natural polymer GT as a traditional medicine have been known from antiquity. Presently, the global market for the GT is estimated to be about 300 tons per year (FAO, 1995). GT has used for centuries as a binder, thickener, emulsifier, stabilizer, and in several fields such as food, beverages, medical, and cosmetic industries. The word Tragacanth (Trag) originated from the Greek words 'tragos' mean goat and 'akantha' mean horn, which describes goat horn like the curved shape of the flakes and ribbons. GT is among one of the best quality of commercial gum. GT has been outlined as: "a dried exudation obtained from the stems and branches of Asiatic species of genus Astragalus, in Leguminosae family" (FAO, 1992b). Genus Astragalus has the largest variability worldwide within the Leguminosae family, including around 2,000 species in 100 sub-divisions (Anderson et al., 1988). Plants develop GT in the sap of the root and seep out readily with injury in the form of 'twisted flakes' or 'ribbons,' which dries out and becomes brittle on air exposure.

Recently, naturally occurring biopolymers, especially GT is gaining increasing attention in the areas of pharmaceuticals and nanocomposite based prolonged drug delivery systems, in special reference to wound healing, tissue regeneration, oral, and topical therapeutic system. The aforementioned applications have unveiled new avenues for Nanomedicine, because specially fabricated GT polymers are capable of delivering bioactive therapeutic substances to the target tissues with controlled specified pharmacodynamics. Particular attention is being paid to GT owing to its biocompatibility, biodegradability, nontoxic, non-immunogenic, an advantage of being readily fractioned and hydrolyzed efficiently, which can subsequently be eradicated as a nontoxic product by metabolic pathways. Furthermore, the GT biopolymers have an excellent amphiphilic architecture, dynamic rheology, high pH, and thermal stability, making it an ideal eco-friendly industrial polysaccharide, mostly employed as a binder, emulsifier, stabilizer, thickener governing widespread applications in food, beverage, cosmetics, and pharmaceutics. Promising avenues of biomedical research have also issued for the versatile controlled drug loading and release approach, making the GT a potential excipient candidate for pharmaceutical industries.

Nonetheless, to our knowledge, the congregation of advanced pharmaceutical implications using naturally occurring carbohydrate polymer GT is obscure. Hence, compiling knowledge gained in recent years is desirable. This chapter delineates the occurrence, extraction, processing, structural composition, physiochemical characterization, safety evaluation, most recent biomedical and pharmaceutical applications and challenges of natural carbohydrate polymer GT, with special focus on varied hydrogel-based tissue engineering, skin repair, nerve regeneration, bone repair, dental repair, nano-encapsulation, and controlled site-specific drug delivery.

3.2 OCCURRENCE AND ROLE IN PLANT

Trag is a natural polysaccharide gum obtained from the several species of Iranian, Turkish, Indian, Afghanistan, and Russian leguminous family of the genus *Astragalus*. Significant diverse species, included, *A. gummifer* Labill., *A. adscendens* Boiss., *A. kurdicus* Boiss., *A. microcephalus* Willd., *A. brachycalyx*, *A. echindnaeformis* Sirjaev., *A. gossypinus* Fisch. Important contributing species are *A. gummifer* Labill., *A. tragacatha* and *A. microcephalus* Willd. These are mostly shrubs of size ranging from 10 cm to 1 m, growing in dry, semiarid regions of the Middle East countries. Exudates gums are produced from the dried sap of *Astragalus* plants to protect the wounds as a defense mechanism to tissue injury. The exudate covers the injured plant part and hardens into flakes once on exposure to air and sunlight (Anderson and Bridgeman, 1985).

3.3   EXTRACTION AND PROCESSING

The natural Trag exudates gum flakes were commercially harvested after inflicting wounds into the plant bark during summer seasons. The curled gum flakes get dried once in contact with air and light. The material is then tapped handpicked, cleaned, and sorted for grading. GT sorted into five to seven grades (FAO, 1995) of quality and powdered for packaging and export. Iran dominates the world market as the biggest producer of high-quality GT (Anderson and Bridgeman, 1985).

3.4   CHEMICAL STRUCTURE AND COMPOSITION

Trag is a plant-derived natural occurring, highly complex polysaccharide, grossly heterogeneous and branched, acidic proteoglycan of high molecular weight of about 840 kDa with 4,500 x 19 Å of elongated shape (Verbeken et al., 2003). As described in Figure 3.1, Tragacanth contains two different fractions, such as Bassorin (water-swellable; 60–70%) and Tragacanthin (water-soluble; 20–30%). Tragacanthin was water-soluble. Tragacanthin is further fractioned into ethanol, insoluble tragacanthic acid and ethanol soluble arabinogalactan. Bassorin and tragacanthic acid are water insolubility, but can form a viscous gel.

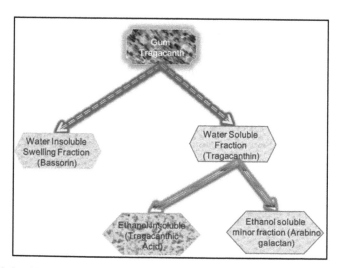

FIGURE 3.1   (See color insert.) Fractional composition of gum tragacanth (GT).

Both fractions of Trag contain small amounts of methoxyl groups and proteinaceous material, the latter present in higher amounts in the water-soluble fraction (Anderson and Bridgeman, 1985) (Figure 3.2).

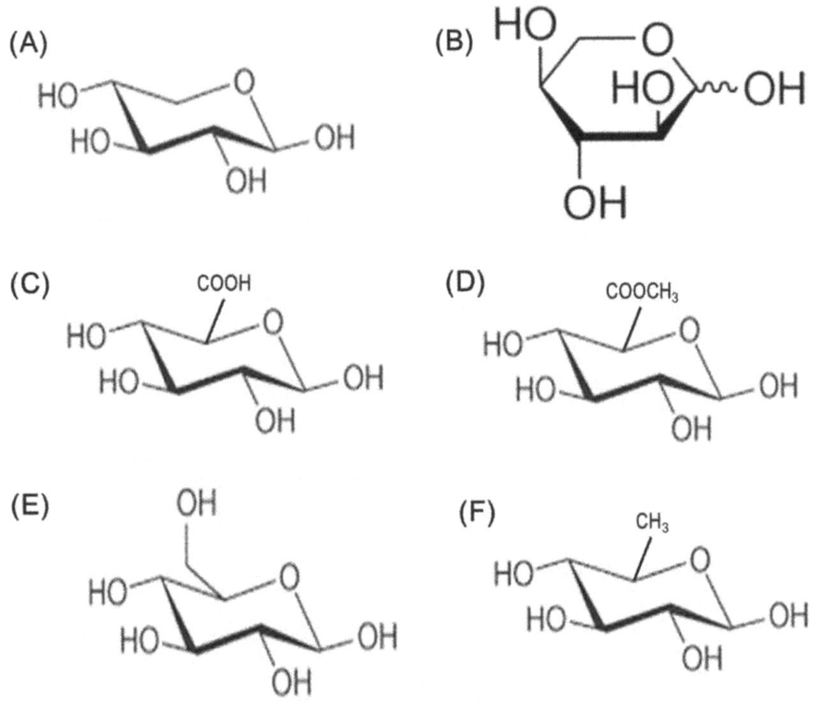

FIGURE 3.2 Partial part of the chemical structure with main chemical building units of gum tragacanth (GT). (a) β-D-xylose, (b) L-arabinose, (c) α-D-galacturonic acid, (d) α-D-galacturonic acid methylester, (e) β-D-galactose, (f) α-L-fucose.

As shown in Table 3.1, gum composition of two most important Turkish shrubs *A. gummifer* and *A. microcephalus* has been illustrated. Fractional composition grossly varies between two different species, even under the same genus and family. Significant differences are observed in methoxyl content, sugar composition, and other proportion of insoluble and soluble components.

So, commercial GT is mostly variable product. There is a variation on its chemical constitution and hydration according to species diversity, geographical variation, growing conditions, environmental impact, and seasonal alteration.

TABLE 3.1 Compositional Data of the GT Exudates Obtained From Turkish *Astragalus* Species[*]

Parameter	A. microcephalus	A. gummifer
Loss on drying (%)	12.7	9.9
Total ash (%)	3.2	2.9
Nitrogen (%)	0.58	0.46
Total protein (%)	3.65	2.84
Methoxyl (%)	3.3	0.9
Ratio of soluble: Insoluble components	65:35	40:60
Galacturonic acid (%)	11	3
Galactose (%)	14	23
Arabinose (%)	37	63
Xylose (%)	22	5
Fucose (%)	12	2
Rhamnose (%)	4	4

[*]Adapted from Anderson and Bridgeman (1985). The sugar composition was assessed and compared after hydrolysis (Reprinted with permission from Verbeken, D., Dierckx, S., & Dewettinck, K., 2003. © Springer Nature.).

3.5 SAFETY EVALUATION

Due to their wide distribution in nature and significant role in food and beverages, cosmetics, textile, and pharmaceutical industries, the extraction, isolation, and characterization of GT are still gaining attention in recent years. The increase in the use of GT as food additives, cosmetics, and in pharmaceutics, also demands its safety evaluation on toxicological and immunological aspects. Transmission electron micrograph of the ultrastructure of rat hearts and livers after dietary supplementation with GT for 91 days showed no detectable abnormalities, and no pathological inclusions in the heart and liver specimens were observed (Anderson et al., 1984). On a clinical trial Eastwood et al., reported GT was well tolerated, and no adverse toxicological effects of breath hydrogen and methane concentrations; plasma biochemistry; glucose tolerance; hematological indices; urinalysis parameters; serum cholesterol, triglycerides, and phospholipids were observed following five male volunteers consumed high dose of 9.9 g GT daily for 21 days (1984).

Hence, GT has been affirmed as safe (GRAS: generally recognized as safe) within the USA since 1961. GT has been permitted as a food additive

(E413), since 1974. A comprehensive study showed that GT did not cause maternal toxicity nor teratogenicity after oral administration (as a suspension in corn oil) at levels up to 1200 mg/kg body weight/day in pregnant mice and hamsters. Mutagenicity and immunogenicity were not observed when GT was evaluated for genetic/allergic activity (Anderson et al., 1989). GT was found to be non-carcinogenic with no significant incidence of the preneoplastic or neoplastic lesion were observed when administered at dietary levels to B6C3F1 mice of either sex for 96 weeks (Hagiwara et al., 1992). In a recent study by Moradi et al., on human serum suggested that GT doesn't induce any adverse change in protein conformation at serum levels (2018). The wide range of research findings summarized indicates that evidence of safety evaluation for GT is substantially appropriate and adequate. Hence, confirming GT as non-immunogenic, non-carcinogenic, well tolerated, overall non-toxic, and safe for pharmaceutical practices.

3.6 PHYSIOCHEMICAL CHARACTERIZATION

GT is the most extensively used natural polymers which have been revolutionized pharmaceutical/biomedical applications recently, and has attractive characteristics like nonimmunogenic, nontoxic nature, biodegradability, bioavailability, higher pH/thermal resistance, anti-inflammatory, nontumorigenic, analgesic, nonallergic and with good antibacterial efficacy (Figure 3.3). The plant *Astragalus* is employed in herbal medicine as a mild laxative. According to folk beliefs of some peoples in the Middle East, Iran, India, and North Africa, it is implied that the leaves contain useful elements in the preparation of poultices and wound healing lotions (Ehsan Fayazzadeh et al., 2014). In modern Iranian medicines, GT also demonstrated the analgesic activity (Seyyed et al., 2015). GT produce through the process of gummosis, exudes out, and dries. Physically, it appears as a few inches long, flat, curled, flexible, opaque, off white or yellowish-brown, odorless, tasteless, brittle plant products.

3.6.1 GELLING

GT swells readily with cold or warm water to form highly viscous semi-gel, even at 1% (w/v). This gum has clarity, good biofilm–forming properties that make stable suspension for the excellent gelling property. It reduces syneresis when blended in combination. The GT aqueous gel stays stable for long up to 15 days that make it an excellent gelling material (Chenlo et al., 2010).

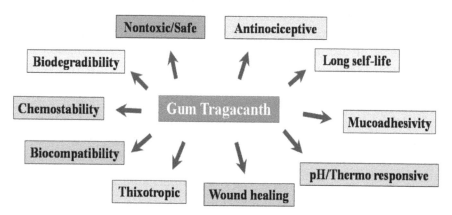

FIGURE 3.3　**(See color insert.)** Characteristic properties of gum tragacanth (GT).

3.6.2 VISCOSITY

Tragacanth Gum (TG) offers a pseudoplastic texture to the solutions. It has excellent thixotropic properties and stable when stored for several days. Aqueous solutions of GT are reckoned as the most highly viscous solutions (approximately 3400 cps) among all the plant gums. Maximum viscosity is attained after heating it for 8 h at 40°C or for 2 h at 50°C or after 24 h at room temperature (Mary, 1987).

3.6.3 SOLUBILITY AND HYDRODYNAMIC STATE

Both the GT fractions of the water-soluble and insoluble part are swellable to form gels. GT is also soluble in alkaline solution and aqueous hydrogen peroxide solution, insoluble in organic solvents, including alcohol. It forms a homogenous solution and emulsion preventing bioactive components from settling down. The hydrodynamic size of GT particles varies with different species, which ranges from 7–38 µm having monomodal distribution (Hassan et al., 2013).

3.6.4 pH STABILITY

GT has excellent acid tolerance properties and is highly stable for a wide range of pH change from 4–8. It is more stable and viscous at pH 5, most unstable at pH 2.2 with reduced viscosity (Mary, 1987).

3.6.5 THERMAL STABILITY

GT is stable over a wide range of temperature of heat and has long shelf-life. GT gives good self-life to its conjugates on temperature fluctuations. Recently, L. Brambilla et al., reported that GT has the same constituents before and after artificial aging. GT has undergone thermal aging consists of 1,000 h of oven heating at 60±5°C, and the FTIR spectra detected no differences in absorption band distribution, confirming that the major contents of GT are stable to thermoxidation (2011).

3.6.6 CHEMICAL STABILITY

TG is compatible with a wide range of chemical formulations. GT is mostly nonreactive and does not deform chemical constituents when blended together. It stabilizes suspensions, and emulsions. It is compatible with various nanoparticle formulations, pharmaceutical drugs, other polysaccharides, electrolytes, carbohydrates, some waxes, proteins, emulsifiers, and detergents (Mary, 1987). Further, Moradi et al., reported that GT has no adverse effect on protein conformation on human serum levels during binding with GT (2018).

3.6.7 ANALGESIC PROPERTY

In traditional medicine, practice species of *Astragalus* were used as an analgesic agent since ancient times. To confirm this, Seyyed Majid Bagheri et al., investigated the antinociceptive effect of GT (varied concentration dissolved in sterilized water) by the rout of intraperitoneal injection of thermal and acetic acid-induced pain using 98 albino male mice. Observed results of the writhing test concluded GT has efficient antinociceptive property compared to standard control diclofenac sodium (DS) and morphine. Most effective concentration GT found to be at 500 μg/kg (2015).

3.6.8   *WOUND HEALING PROPERTY*

In Indian Ayurveda, Chinese, and Iranian herbal medicine GT has been used as a wound-healing agent. Abdolhossein Moghbel et al., contemplated the effect of topical application of GT on excisional wounds in male Iranian rabbits and suggested that Trag mucilage to be a more potent wound healing agent and could significantly give the best healing effect at 6% w/v GT cream. This treatment exerted the lowest period of wound closure and healing compared to standard control Eucerin and commercial 1% Phenytoin cream (2005). In addition, Fayazzadeh et al., prepared gel of Trag at 5% (w/v) by dissolving dried GT powder into sterilized water that was naturally obtained from plant *Astragalus gummifer* and investigated in full-thickness excisional skin wounds healing in albino Wistar male rat. GT-treated wounds were almost completely closed by the 10th day of the treatment. Observed results reasoned out that, components of GT have fastened the transition phase of the inflammation and tissue granulation on the healing process of the wound and meliorated wound healing index, and extracellular matrix (ECM) remodeling which quickened the wound contraction and closure. This study demonstrates the efficacy of GT as an effective, biocompatible, inexpensive, easy to use, and readily available product for wound dressing. Further suggesting GT-containing formulations would be effective preparations for the treatment of chronic wounds such as diabetic foot ulcers, ischemic wounds, burn wounds, and bedsores (2014).

3.7   RHEOLOGY

Rheology describes 'flow behavior' of liquid, soft solid or solid in response to applied stress/force, that can be described by plotting shear diagram of shear rate versus shear stress. S. Balaghi et al., investigated flow behavior of six GT species of Iran and the results show that all species of GT dispersions had rheology of shear-thinning nature (2010). In rheology, shear thinning means the non-Newtonian fluid behavior where the viscosity decreases with an increase in shear strain. GT possesses pseudoplastic behavior, making it thixotropic, which on withdrawal of strain reverses back to its original state. Rheological behavior of GT varies with a diversity of species as differences in its constituents. While 1% Bassorin solution at 25°C shows a comparatively higher viscosity gel-like texture, Tragacanthin solution behaves like complex coil polymers of concentrated solution. However, little deformation in rheological property has been observed between Bassorin and Tragacanthin. The viscosity of both the fractions will decrease by increasing the temperature, but the viscosity of the Bassorin is less sensitive to heat than Tragacanthin. Moreover, for similar concentrations, Bassorin solution has a higher viscosity than Tragacanthin and whole GT. This confirms GT fractions have differential rheological properties within the same species (Mohammadifar et al., 2006).

3.8 RECENT ADVANCES IN PHARMACEUTICAL APPLICATIONS

TG is one of the oldest traditional pharmaceutical ingredients known, being described by Theophrastus several centuries (Strobel et al., 1986). In Indian, Ayurveda GT has been used extensively for medicinal purposed since antiquity. Nano-formulations such as hydrogels, liposomes, micelles, and conjugates have shown advances in drug delivery and gene delivery. Gum has also been used in denture adhesives, poultices, and in a quantitative fluorescent antibody assay of polioviruses (Kedmi et al., 1978). Dietary consumption, topical application of Trag is considered safe and non-toxic as reported by several studies using animal and human experiments (Moradi et al., (2018). Recent advances in the GT application in pharmaceutical research summarized in Table 3.2. Encapsulation is a technique of packing bioactive molecules such as pharmaceuticals, genes, growth factors, and live cells to protect from the external environment. Notably, advantages of micro/nano-capsulation are the protection of bioactive substance, multi-stimuli sensitivity, controlled release, and site-specific release of the bioactive core material with higher efficiency. Tissue engineering, natural polymer scaffolds provide a three-dimensional texture for cell adherence, proliferation, and differentiation, nutrient, and metabolite circulation of nutrient and cell metabolite as well for tissue regeneration. Natural polymer hydrogels have been gaining immense attention as a scaffolding material for varied tissue regeneration. As GT possesses nontoxic, antibacterial, antiviral, antinociceptive properties as well as nonimmunogenic, inexpensive, and readily available, been extensively used in biomedical and pharmaceutical settings. GT-based nanoformulations with the desired bioactive drug were proven excellent drug release platforms in various pathophysiological experiments such as wound healing, cancer treatment in animal models, and clinical settings (Ghayempour and Montazer, 2017a, 2018).

Based on nanobiotechnology methods, GT-based nano-formulations have been profiting significant advances in areas such as tissue engineering, drug delivery, and pharmacotherapeutics. As evident from several reports discussed in this chapter, the spectacular improvement has been achieved with carbohydrate polymer GT compounds having indubitable benefits for pharmaceutics, used as nanoparticles, cryogel, xerogel, nanobeads, nanocapsule, nanohydrogel, nanocomposite, nanofiber, nanocargo, nano scaffold, nanomicelle for drug and gene delivery system (Figure 3.4).

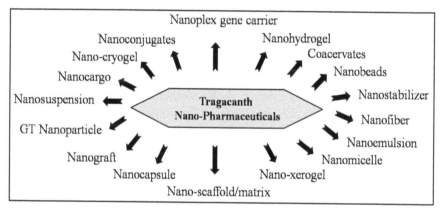

FIGURE 3.4   Schematic of possible nano-biopharmaceutical applications of gum tragacanth (GT).

Recently, GT has been used in an overwhelming number of hydrogel-based nano-pharmaceutical applications. Slow and sustained release is highly desirable in the case of therapeutics that demands prolonged administration of the drug. The GT-nanocomposite hydrogel, nanogel, xerogel, and cryogel possess high porous/fibrous network structure, increased surface area, non-toxicity, swelling/de-swelling behavior and have been extensively studied in biopharmaceutical applications as a tissue engineering biomaterial, drug encapsulation, and drug delivery. In combinations with other polymers, GT provides good mechanical properties, which provides physical and biocompatible scaffold support for cell adhesion and regeneration as well as for excellent proliferation. There is an increasing demand for nanoparticle based-hydrogel with, multi-stimuli responsiveness, biodegradability, bioadhesivity, standard diffusion coefficient, swellability as well as multifunctionalities with enhanced physic-mechanical properties for the development in the rapid formulation of Nanomedicine. As described in Figure 3.5, various controlled stimuli such as pH, infrared laser, magnetic or electrical signals used as Microswitch to trigger GT-based nanocomposite or nanosystem encapsulated versatile active ingredients, to release drugs to the specific organ/tissue. Drugs releases on slow/fast mode upon requirement in a controlled/regulated manner to the target site without harming neighboring normal tissues or organs. Sustained prolonged, localized, and on-demand release of drugs with reduced cytotoxicity has dramatically improved drug efficacy that has emerged as an advanced practice in the new era of pharmaceuticals.

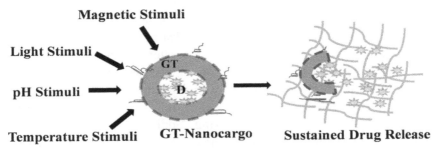

FIGURE 3.5 **(See color insert.)** Schematic of multi-stimuli responsive GT-encapsulated/ hydrogel based site-specific drug delivery system (GT – Gum Tragacanth, D – desired drug).

GT nanocomposite hydrogel could be an appropriate cargo to load and transport therapeutic biochemical factors such as drugs, growth factors, etc. Despite extensive use of biopolymers in biomedical applications, its long-term interaction mechanism remains as a challenge for clinical acceptance and approval. This opens a new avenue to develop new methodology and techniques for the detailed interaction mechanism study as an emerging field on vast potential GT-based nano-pharmaceutical and drug discovery.

In spite of the availability of synthetic alternative materials, use of natural biopolymer GT has been increasingly demanding in pharmaceutical and biomedical sectors owing to its unique physicochemical properties (swellability, viscosity, safety, nonreactive, inexpensive, readily available, emulsification) combined with a high degree of pH/thermal/chemical stability. Following are the pharmaceutical applications described in detail;

3.8.1   SUSTAINED RELEASE TABLET FORMULATIONS

Sustained release tablets have advantages over conventional one and been highly demanding of pharmaceutical industries. Several reports using GT as sustained release formulations been popular recently. Akhgari et al., prepared a tablet formulation, taking GT in different ratio with acacia and model drug ibuprofen and theophylline at concentrations of 20%, 40%, and 60%. This tablet formulation showed the extended release of model drugs that can be used successfully as binders in the sustained release pellet formulations (2011). In another GT tablet formulation, dipyridamole (DIP) alginate (Alg), Trag tabletted microspheres were prepared by micro-encapsulation method, emulsification method at ratio DIP: alginate: GT in (1:1:1) and found to be prolonged dug release with good mechanical strength (Gursoy et al., 1999). Formulations where the polyvinyl pyrrol-idone (PVP K90) was partially mixed with GT also sustained the release of DS drug for 12 hours following Higuchi model and Non-Fickian diffusion in human (Iqbal et al., 2011). Shahiwala et al., developed sustained release matrix tablets containing DS using Trag, Sodium Alginate (SA), and Gum Acacia (GA) natural polymers. The results showed the highest drug release profile of GT in the form of extended release action in the order of: GT > SA > GA in vivo as similar to marketed Diclofenac tablet (2018). Polymer matrix xanthan gum and Trag incorporated by matrix compression technique for release of metoprolol tartrate drug. The results showed that the formulation containing 30% xanthan gum and 10% GT enhanced oral sustained-release tablets of metoprolol tartrate is of marketing standard (Rasul et al., 2010).

3.8.2   EMULSION STABILIZER

GT has been used in several emulsions, suspension as a reducing and stabilizing agent. Incorporation od GT stabilized *in-situ* synthesis of zinc oxide nanoparticles using an ultrasound irradiation method that showed a better photocatalytic activity and 100% antimicrobial properties against *Staphylococcus aureus, Escherichia coli* and *Candida albicans* (Ghayempour et al., 2017b). It was reported that the deoxyhexoxyl groups or themethoxyl groups in the GT structure might be a reason for emulsion stabilization. Notably, the total content of galacturonic acid, methoxyl groups in the GT might play a key role in TGs to stabilize varied emulsions (Hassan et al., 2013). The phenomenon GT as a stabilizer also relates to the negatively charged carboxylic groups in galacturonic acid that is the main backbone of fraction tragacanthin. GT has zeta potential around −21 mV that adds to the higher stabilizing property of a different emulsion system (Azarikia et al., 2010). Fabrication of GT stabilized green Au NPs, as a carrier for Naringin, can act as effective drug delivery vehicles with the added antibacterial efficacy of Naringin (Rao et al., 2017).

3.8.3 COACERVATES

A complex coacervate is the process of coating, pharmaceutical agents with cationic and anionic interactions of the polymer matrix in water. Beta-carotene microcapsules were coated with complex coacervates of casein/gum tragacanth (CAS/GT). This process of coacervation using GT made it stable for 3 months under three temperature levels of (5, 25 and 40 degrees C). This encapsulation process strongly increased the stability of micronutrients inside the GT coacervates and effective for drug delivery and sustained release (Jain et al., 2016).

3.8.4 ANTIVIRAL AGENT

Several above mentioned reports indicated towards antimicrobial properties of GT. Tragacanthin which is a fraction of GT polysaccharides extracted from *Astragalus brachycentrus* (AV208) injected to Punta Toro virus (PTV) exposed mice at 1.6–50 mg/kg/day 24 h post exposure of virus, reduced mortality, liver, and spleen virus titers, liver infection scores and serum transaminases. This report suggests Tragacanthin can be useful for the treatment of Bunyaviruses and Hantavirus infections in human (Smee et al., 1996).

3.8.5 SUSTAINED DRUG RELEASE POLYMER MATRIX

The polysaccharide of GT provides an outstanding polymer network for adsorbing desired drugs. Its porous structure helps to the prolonged slow release of drugs upon stimuli microswitch control in a cell-specific manner. New pH/temperature controlled GT graft copolymers were synthesized by click chemistry. It formed micelle and used as quercetin (QC) loaded nanocarriers. Drug QC was released in a sustained and prolonged way following first-order model (Hemmati et al., 2016). Another novel drug delivery nanocomposite was prepared using GT, polylactic glycolic acid (PLGA) and tetracycline hydrochloride (TCH), which showed sustained drug release for 75 days along with increased biocompatibility and anti-bacterial action. This makes GT-based PLGA-TCH nanofibers composite a potential candidate for the delivery system in periodontal diseases (Ranjbar-Mohammadi et al., 2016).

3.8.6 ANTI-ADHERENT/ANTI-CARIES AGENT

Acidogenic bacteria *Streptococcus mutans* adhere to dental enamel live in forming biofilm and erodes dental enamel with acid production resulting into a caries. GT at 0.1% concentration in water and mouthwash reduced biofilm formation and significantly decreased *S. mutans* adherence to the enamel on the surface pretreatment of hydroxyapatite in rats and in human preventing dental plaque formation (Shimotoyodome et al., 2006).

3.8.7 SKIN SUBSTITUTE

GT, along with poly (epsilon-caprolactone) (PCL), nanofibers patch were prepared through the electrospinning method. This wound-healing mat portrayed outstanding wound care efficacy of GT with enhanced cell adhesion, growth, and proliferation of fibroblast. This biocomposite showed antibacterial potential against *Staphylococcus aureus* and *Pseudomonas aeruginosa*, making it an excellent skin substitute and a potential candidate for skin tissue engineering (Ranjbar-Mohammadi et al., 2015).

3.8.8 NANOPARTICLES AND NANOPLEXES

Nanoparticles are particles which are in nano-range typically the size range from 1–100 nm having unique physicochemical properties and high surface area to volume ratio. Nanoplex is a core-shell type nanoparticle, which enacts nucleic acids like DNA or RNA for target delivery as therapeutics. The nanoparticles of GT-oligochitosan used to make nanoplexes with the incorporation of DNA via electrostatic interaction process. This nanoplex as novel gene carriers have been investigated for transfection efficiency on Hela and HepG2 cell lines. Obtained results indicated a high transfection potential as compared to chitosan polyplex (Fattahi et al., 2013). Sadrjavadi et al., prepared de-esterified tragacanth-Chitosan nanoparticles (DET-CS NPs) as a novel carrier for methotrexate and optimized using Taguchi design. In vitro sustained release was observed within nine days. The obtained results suggest that DET-CS NPs a suitable candidate for targeted methotrexate delivery to cells containing asialoglycoprotein receptors (ASGPR) (2018).

3.8.9 MUCOADHESIVE

Mucoadhesive are bioactive drug delivery pharmaceutics (tablet, patch, emulsion), which adheres to the mucosal surface of the gastrointestinal tract (GIT), oral cavity, nasal cavity, vaginal cavity and mucosae surrounding the eyes for sustained release of the dosage via mucosal epithelia, cilia. Over the past few decades, GT-based mucosal drug delivery has received great attention for prolonged, controlled drug release improving therapeutic outcome. Recently, Jabri et al., (2018) used GT composed gold (AuNPs) nanoparticle for loading Amphotericin B (AmpB) through the solvent diffusion method and investigated for enhanced oral bioavailability in a rabbit model. Results revealed that GT-Au-AmpB hybrid NPs enhanced AmpB oral bioavailability in a sustained and controlled manner (Jabri et al., 2018).

3.8.10 WOUND HEALING

Advanced wound healing products contain ointments, dressing pads, bandages, wound closure tapes, hydrogels, etc., which provides a moist and occlusive environment in the complex cellular event of tissue regeneration and restoration with added antibacterial effects. GT itself reported to be non-toxic, antibacterial, and hygroscopic; interestingly when used in combinations it enhances the potentials of the wound healing process. A GT-based wound-healing product was prepared by entrapping aqueous *Aloe Vera* extract into the GT polymer wall. This wound care product improved human fibroblast cell growth, proliferation (98%) and migration with relative high antimicrobial activities against *E. coli, S. aureus,* and *C. albicans* (Ghayempour et al., 2016a). Further, another study has demonstrated curcumin (Cur)-loaded PCL/GT membranes, which provided the controlled release of curcumin for over 20 days, making it a potential wound care dressing patch (Ranjbar-Mohammadi et al., 2016).

3.8.11 *MICRO/NANOCAPSULE*

Micro- and nanocapsules are capsules of a micrometer or nanometer range used for biomedical and pharmaceutical applications as a drug carrier. Trag nanocapsules are containing Chamomile extract prepared through sono-assisted microemulsion method and UV cured onto cotton fabrics. This GT-based nanocapsule showed Chamomile release up to 96 h providing a high stability as well as a good antimicrobial activity up to 94% against *S. aureus, E. coli* and *C. albicans* (Ghayempour et al., 2017c). In addition, TG as a natural polymeric wall for producing antimicrobial nanocapsules loaded with peppermint oil showed a 100% microbial reduction after 12h against *Escherichia coli, Staphylococcus aureus* and *Candida albicans* (Ghayempour et al., 2015).

3.8.12 *PROBIOTICS*

Probiotics are the edible ingredients that improve gut microbiota by facilitating growth and activity of beneficial bacteria such as *Lactobacillus sp.* and *Bifidobacteria species.* Gavlighi et al., investigated that enzymatic depolymerized GT increased the growth of different species of *bifidobacteria* and *Lactobacillus* as well as decreasing growth of harmful pathogenic bacteria *Clostridium perfringens* (2013). This study suggests the potential of another exciting probiotic of GT that can be exploited in pharmaceutical industries.

3.8.13   SPECIALIZED BIOMATERIAL/TISSUE ENGINEERING SCAFFOLD

Scaffolds are typically made up of polymeric biomaterials with 3D texture to orchestrate cell adhesion, nutrient flow, and tissue regeneration. Different kinds of biomaterial scaffolds have been prepared to execute desired functions. GT hydrogel-based scaffold used for osteogenic differentiation of human adipose-derived mesenchymal stem cells displayed the highest proliferation on day 5 and nontoxic with highest ALP activity, osteogenic gene expression, and mineralization when compared to other polymer-based scaffolds. These observations demonstrate that GT hydrogel-based scaffold could be appropriate for acceleration and differentiation of osteogenic stem cells and can have potential orthopedic applications (Haeri et al., 2016). Another group prepared gum tragacanth-based poly (l-lactic acid) (GT/PLLA) nanoscaffold at 25:75 compositional ratio by electrospinning method. GT/PLLA scaffold acted favorable to support nerve cell growth and elongation as compared to PLLA alone. This shows PLLA/GT 75:25 compositional ratios are promising substrates for the application of neuronal tissue regeneration as bioengineered grafts (Ranjbar-Mohammadi et al., 2016).

3.8.14   NANOFIBER

Nanofibers are a fiber of nanometer range possessing unique physico-chemical properties. Natural polymer nanofibers are now being an excellent candidate for a number of biomedical and pharmaceutical applications such as in medicine, biotechnology, and tissue engineering, owing to its large surface area to volume ratio and nanostructure. Fabricated GT-nanofibers of 50nm has shown hydrogel properties of swelling of 142% and drying time of 3h, making it an excellent biomaterial for tissue and organ regeneration (Ghayempour and Montazer, 2018). In another formulation, GT, and poly (vinyl alcohol) (PVA) nanofibers were produced using the electrospinning technique. This nanofiber showed good human fibroblast growth and proliferation as well as enhanced antibacterial property against *Staphylococcus aureus* and *Pseudomonas aeruginosa*, making it an excellent wound dressing (Ranjbar-Mohammadi et al., 2013). Zarekhalili et al., (2017) prepared GT-based PCL and PVA complex nanofiber by two nozzles electrospinning process. This nanofiber scaffolds investigated for skin tissue engineering and resulted in 95.19% antibacterial against *S. aureus* in addition to enhanced fibroblast proliferating suggesting GT a better candidate in 3D scaffold preparation for tissue regeneration and skin substitutes (Zarekhalili et al., 2017).

3.8.15 MICROENCAPSULATION

Microcapsules are small spheres containing tiny materials like drugs, cells, gene, etc. Recently biodegradable natural polymer GT has been used to make microcapsules encapsulating bioactive drugs for better, controlled, and site-specific delivery of desired drugs. GT matrix-core shell microcapsules encapsulated live βTC3 cells following the Taguchi method and resulted in better stability up to day six with increased cell adhesion, increasing cell viability (Alvandimanesh et al., 2017). Asghari-Varzaneh et al., reported that iron salt ($FeSO_4.7H_2O$) encapsulated in GT hydrogel microsphere using a solvent evaporation method reached to target site and increased duodenal absorption of iron on pH response (2017). This proves microencapsulation is an advanced method and can revolutionize pharmaceuticals.

3.8.16 HYDROGELS

Hydrogels are hydrophilic swellable polymeric materials with three-dimensional porous structure. These are the first biomaterials prepared for use in human biomedical studies (Salamanca et al., 2018). Synthesis of hydrogels includes cross-linking a group/single hydrophilic polymeric material. There are different kinds of hydrogels available, like self-healing, self-assembling, injectable hydrogels, etc. Injectable GT-hydrogels were prepared by enzymatic oxidation process taking GT-tyramine conjugate. The release profile of drug from hydrogels followed Higuchi and Fickian diffusion mechanism. Finally, it was evinced that the GT has bettered the swelling characteristics and drug release behavior of this polymeric network (Dehghan-Niri et al., 2015). In another hydrogel formulation was synthesized using TG-cl- poly (lactic acid-co-itaconic acid) through graft copolymerization reaction by microwave-assisted technique and loaded with Amoxicillin exhibited the highest Amoxicillin. This GT based hydrogel showed drug release of about 96% at pH 2.2 after 6h in addition to mild antioxidant properties, and 43.85% of free radical scavenging activity was at a concentration of 640mug/mL and effective antibacterial property against *S. aureus* (Gupta et al., 2018). In addition, calcium carbonate nanoparticles (CC NPs) and TG hydrogel nanocomposite (HNC) were highly effective for significant heavy metal removal such as Pb ($^{2+}$) ions from water (Mallakpour et al., 2018). Further, polyacrylonitrile and Iranian TG grafted to form biosorbent nanohydrogel, which selectively removed heavy metals like Co, Zn, Cr, and Cd from water (Masoumi and Ghaemy, 2014). Recently, Mohamed et al., investigated AmB, Trag, and acrylic acid (AAc) – (Trag/AAc)-AmB hydrogel antifungal efficacy in a systemic candidiasis mice model. Results are shown increased oral bioavailability and potent antifungal efficacy as evidenced by prolonged survival time and low tissue fungal burden, as well as decreased inflammatory cytokines in liver and kidney tissues (2018). Another TG/nanosilver hydrogel coated cotton fabric showed better water absorption properties and antibacterial properties against Escherichia coli and Staphylococcus aureus (Montazer et al., 2016). Nur and Vasiljevic reported insulin-Trag hydrogel improved insulin entrapping and increased oral availability of insulin (2018). Another mucoadhesive hydrogel prepared by cross-linking of poly (vinylpyrrolidone)/AAc with TG has shown pH driven ciprofloxacin release following non-Fickian diffusion mechanism (Singh and Sharma, 2017). Wound healing hydrogel prepared using PVA, TG, and SA entrapped with the antibacterial moxifloxacin drug through radiation method. Drug release profile followed non-Fickian mechanism with thrombogenicity (82.43+/–1.54%), making the hydrogel best fitted for wound dressing applications (Singh et al., 2016).

3.8.17 SPHERICAL HYDROGEL BEADS

In an experiment, Rahmani et al., prepared spherical porous hydrogel beads based on ion-cross-linked GT and graphene oxide (GO) and found hydrogel beads released drug Rivastigmine under pH simulated gastric (<45% at pH 1.2) and intestinal (approximately 97% at pH 7.4) media. Further, GT increased the swelling efficiency, entrapment capacity, and a pH-controlled release of the Rivastigmine drug (2018).

3.8.18 NANOGELS

A nanogel is a biopolymer cross-linked nanoparticles hydrogel, which provides an increased surface area in its porous polymer matrix for the storage of versatile biomolecules. GT-based nanogel composed of super-paramagnetic Fe_3O_4/SiO_2 nanoparticles, N-vinyl imidazole (VI) and template QC prepared by a sol-gel process. This nanogel provided excellent electromagnetic property to absorb the template QC drug quickly with the maximum capacity of 175.43 mg g (−1) following the Langmuir model. This GT-based nanogel has excellent drug recognition and binding affinity, making it an advanced tool for drug delivery system with wide application in nano-pharmaceuticals (Hemmati et al., 2016).

3.8.19 CRYOGEL/XEROGEL

Super macroporous cryogel of silymarin (SM) loaded gum tragacanth-poly (vinyl alcohol) (GT-PVA) was prepared at sub-zero temperature, and xerogels were prepared which improved mucoadhesivity and oral bioavailability of SM significantly (Niknia and Kadkhodaee, 2017). Detailed pharmaceutical applications of GT summarized in Table 3.2.

3.9 CHALLENGES AND LIMITATIONS

Chemical composition of GT greatly diverges with the multifariousness of species, which is a major challenge to maintain pharmaceuticals confidence and export quality. Around 2,000 species diversity of the *Astragalus* genus of having largest Leguminosae family is one of the major drawbacks for Trag recovery with a stable composition (FAO, 1995). The carbohydrate and polypeptide content of GT attracts microbial contamination upon exposure to moisture. Nevertheless, the rate of microbial degradation is slow. To prevent contamination, it mandates proper preservation. There is a variation on its chemical constitution and hydration concording to growing conditions, seasonal variation, and environmental impact. Further, geological variation and species diversity add limits to its physiochemical properties. Longer storage time reduces its viscosity. Recently, GT has been used in a vast number of hydrogel-based nano-pharmaceutical

TABLE 3.2 List of Gum Tragacanth (GT) Applications in Pharmaceuticals

Types	Modification/Formulation	Active ingredient	Application	Reference
Tablet	Wet granulation method, PVP K90 in combination with gum tragacanth	Diclofenac sodium	Higuchi model and super case-II and Non-Fickian diffusion with sustained the release of drug for 12 hours in vivo, in vitro, clinical trial	Iqbal et al., 2011
Tablet	Polymer matrix xanthan gum and tragacanth incorporated by matrix compression technique.	Metoprolol tartrate	Oral sustained release tablets of metoprolol as per marketed preparation	Rasul et al., 2010
Tablet	Sustained release matrix tablets containing diclofenac sodium prepared with Gum Tragacanth	Diclofenac sodium	Drug release profiles were better compared to other natural polymers with the market standard of sustained release of diclofenac	Shahiwala and Zarar, 2018
Pellet	Extrusion-spheronization Acacia and tragacanth (8:2, 9:1, and 10:0 ratio	Ibuprofen and Theophylline	Higher mechanical strength and a faster drug release rate	Akhgari et al., 2011
Stabilizer	GT stabilizes Gold nanoparticles (AuNPs) as a carrier for Naringin,	Naringin	The results suggest that GT stabilized green AuNPs can act as effective delivery vehicles for enhancing bactericidal potentials of Naringin.	Rao et al., 2017
Denture Adhesive	Tragacanth gum paste	GT	Denture adhesives seem useful to depict the fate of denture adhesives over a period of time.	Koppang et al., 1995
Denture anti-adhesive agent	0.1% GT in water and in mouthwash in hydroxyapatite (HA) dental surface pretreatment	GT, Hydroxyapatite	Significantly reduced saliva-promoted adhesion of Streptococcus. mutans MT8148 and dental biofilm formation on enamel.	Shimotoyodome et al., 2006

TABLE 3.2 *(Continued)*

Types	Modification/Formulation	Active ingredient	Application	Reference
Drug delivery micelle	Ring-opening polymerization (ROP) and atom transfer radical polymerization (ATRP), azide-functionalized GT. probe sonication method	Quercetin	First order model thermo/pH-sensitive drug delivery	Hemmati and Ghaemy, 2016
Tissue engineering scaffold	Poly (l-lactic acid) and gum tragacanth (PLLA/GT) nanofibers scaffold prepared by electrospinning	PLLA/GT 75:25 ratio	PLLA/GT 75:25 nanofibers are promising substrates for application as bioengineered grafts for nerve tissue regeneration.	Ranjbar-Mohammadi et al., 2016
Drug delivery system	Via blend electrospinning and coaxial electrospinning polylactic glycolic acid (PLGA), gum tragacanth (GT) and tetracycline hydrochloride (TCH) nanofiber drug carrier prepared	Tetracycline hydrochloride (TCH)	Drug release was sustained for 75 days. Best suitable for periodontal diseases	Ranjbar-Mohammadi et al., 2016
Sustained release tabletted microspheres	Microencapsulation method, Needle extrusion method, Emulsification method DIP: alginate: GT in (1:1:1) ratio used	Dipyridamole (DIP)	Good physical properties, prolong DIP release.	Gursoy et al., 1999
Nanoplexes gene carrier	Electrostatic interactions method of DNA and oligochitosan-tragacanth nanoparticles	GT-DNA-nanoplex	Improved transfection efficiency into Hela and HepG2 cells	Fattahi et al., 2013
Mucoadhesive hybrid nano-carrier	AmpB loaded lecithin NPs hybrid with GT gold NPs prepared by solvent diffusion method	Amphotericin B (AmpB)	Hybrid NPs as efficient carriers for enhancing AmpB oral bioavailability in a controlled manner in the rabbit model	Jabri et al., 2018

TABLE 3.2 *(Continued)*

Types	Modification/Formulation	Active ingredient	Application	Reference
Coacervate	Coacervate of casein/gum tragacanth (CAS/GT) encapsulating beta-carotene	Beta-carotene	Beta-carotene micronutrient was stable for 30 days with sustained release.	Jain et al., 2016
Hydrogel	Enzymatic oxidation of gum tragacanth-tyramine conjugate	Bovine serum albumin and insulin.	Improved drug release behavior	Dehghan-Niri et al., 2015.
Hydrogel	Graft copolymerization reaction using the microwave for tragacanth gum-cl- poly (lactic acid-co-itaconic acid)	Amoxicillin	Amoxicillin loading (73%), drug release of about 96% at pH 2.2=after 6h, antioxidant, antimicrobial	Gupta et al., 2018
Hydrogel tissue engineering scaffold	GT hydrogel prepared by suspension method and was treated with NaOH to obtain pH 7 in order to seed adipose-derived mesenchymal stem cells	GT	Scaffold for accelerating and supporting the adhesion, proliferation, and osteogenic differentiation of stem cells	Haeri et al., 2016
Hydrogel	Amphotericin B (AmB), a pH-sensitive drug carrier composed of Tragacanth (Trag) and acrylic acid (AAc) (Trag/AAc)-AmB-hydrogel prepare by gamma-irradiation method	Amphotericin B (AmB)	Oral bioavailability of Amphotericin B (AmB) increased significantly the antifungal activity in a mouse model of systemic candidiasis	Mohamed et al., 2018
Hydrogel	Tragacanth gum/nanosilver hydrogel coated cotton fabric	GT+nano silver	Hydrogel treated cotton fabrics showed good water absorption properties and anti-bacterial effectiveness against Escherichia coli and Staphylococcus aureus	Montazer et al., 2016

TABLE 3.2 *(Continued)*

Types	Modification/Formulation	Active ingredient	Application	Reference
Hydrogel	Inclusion of insulin into a tragacanth hydrogel by physical complexation between polyelectrolytes	Insulin	Tragacanth-insulin hydrogel increased oral bioavailability of insulin.	Nur and Vasiljevic, 2018
Hydrogel	Ciprofloxacin loaded TG based hydrogels polymerization with polyvinyl pyrrolidone was prepared	Ciprofloxacin	GT-based mucoadhesive hydrogel shown ph driven ciprofloxacin release following non-Fickian diffusion mechanism	Singh and Sharma, 2017
Spherical hydrogel beads	GT and graphene oxide (GO) hydrogel beads prepared using Ca (+2) and Ba (+2) ions and calcium carbonate particles as a solid porogen,	Rivastigmine	pH-sensitive drug release under simulated gastric (<45% at pH 1.2) and intestinal (approximately 97% at pH 7.4)	Rahmani et al., 2018
Nanogel	By sol-gel process using GT crosslinker, Fe3O4/SiO2 nanoparticles, and N-vinyl imidazole (VI) functional monomer in the presence of template Quercetin	Quercetin	Quick adsorption of the drug following the Langmuir model with the maximum capacity of 175.43 mg g (−1) with excellent recognition and binding affinity toward QC and prolong drug release.	Hemmati et al., 2016
Nanohydrogel	GT, 3-aminopropyltriethoxysilane (APTES) and glyceroldigly-cidylether (GDE), poly (vinyl alcohol) (PVA), and glutaraldehyde (GA)	Indomethacin (IND)	*In vitro* IND release was in the range of 50–80% at pH 9 after 24h.	Sadat et al., 2016
Nanohydrogel	De-esterified tragacanth-chitosan nanoparticles (DET-CS NPs) were processed using Taguchi design incorporating methotrexate	Methotrexate	Enhanced drug incorporation and Nanoparticles endocytosed via asialoglycoprotein receptors with sustained release of drug for nine days	Sadrjavadi et al., 2018

TABLE 3.2 *(Continued)*

Types	Modification/Formulation	Active ingredient	Application	Reference
Xerogel/Cryogel	Cryo- and xerogels of silymarin (SM) loaded gum tragacanth-poly (vinyl alcohol) (GT-PVA) were prepared	Silymarin (SM)	Improved mucoadhesivity and oral bioavailability of SM	Niknia and Kadkhodaee, 2017
Skin substitute nanofiber mats	Gum tragacanth (GT) and poly (epsilon-caprolactone) (PCL) nanofiber mat was prepared by solvent diffusion and electrospinning method	GT	Antibacterial, wound healing, skin regeneration.	Ranjbar-Mohammadi et al., 2015
Nanofiber scaffold	Poly (epsilon-caprolactone) (PCL), poly (vinyl alcohol) (PVA) and gum tragacanth (GT) complex nanofiber prepared by two nozzles electrospinning process	GT	Scaffolds prepared are suitable for skin tissue engineering with 95.19% antibacterial against S. aureus and enhanced fibroblast proliferating	Zarekhalili et al., 2017
Matrix-core shell microcapsule	Co-flow extrusion method by the Taguchi design	BetaTC3 pancreatic cells	Mass transfer, cell adhesion, Cell proliferation, and tissue regeneration	Alvandimanesh et al., 2017
Nanocapsule	Sono-assisted W/O/W microemulsion and UV cured on cotton fiber	GT-Chamomile extract	Good release behavior of 96h, a high stability, antimicrobial	Ghayempour and Montazer, 2017a
Nanocapsule micelle	Ultrasonication, microemulsion method	GT-peppermint oil	100% microbial reduction	Ghayempour et al., 2015
Reducing and stabilizing agent	1%GT + zinc nitrate hexahydrate NaOH, sonosynthesis method	ZnO nanorod	100% antimicrobial properties against Staphylococcus aureus, Escherichia coli and Candida albicans.	Ghayempour and Montazer, 2017b

TABLE 3.2 *(Continued)*

Types	Modification/Formulation	Active ingredient	Application	Reference
Nanofiber	Microemulsion method for hydrogel Tragacanth nanofibers	GT	High swellability nanomatrix hydrogel for tissue engineering	Ghayempour and Montazer, 2018
Cell culture media	2.5%, Tragacanth Gum solution (2.5 g/100 ml 0.0t N NaOH) (Tragacanth-FA method)	Tragacanth	Rapid quantitative fluorescent antibody assay of polioviruses	Kedmi, 1978
Antinociceptive agent	GT500 mug/kg Intraperitoneal injection	GT	Antinociceptive effect through the adrenergic system	Bagheri et al., 2015
Antiviral agent	Tragacanthin injected to Punta Toro virus (PTV) exposed mice at 1.6–50 mg/kg/day 24 h post exposure of virus	GT	Reduced mortality, liver, and spleen virus titers, liver infection scores, and serum transaminases. This reports Tragacanthin can be useful for the treatment of bunyaviruses and hantaviruses infections in human	Smee et al., 1996
Nanocomposite fiber	Electrospinning and blended nanofiber of GT+ nanoclay+PVA	GT	Tissue engineering and wound healing	Heydary et al., 2015
Microencapsulation	Solvent evaporation method	Iron salt (FeSO4.7H2O)	Absorption in duodenum.	Asghari-Varzaneh et al., 2017
Suspension	500 mug/kg IP	GT	Antinociceptive activity on thermal and acetic acid-induced pain in mice	Bagheri et al., 2015
Nanoparticles	1%GT + zinc nitrate hexahydrate NaOH, sonosynthesis method	ZnO nanorod	Reducing and stabilizing agent, Non-cytotoxic, strong Antifungal/antibacterial	Ghayempour and Montazer, 2017c
Prebiotic	Enzymatic depolymerization	GT	Enhanced growth of Bifidobacterium, Lactobacillus	Gavlighi et al., 2013

TABLE 3.2 *(Continued)*

Types	Modification/Formulation	Active ingredient	Application	Reference
Wound healing paste	Topical application	GT	Acceleration of skin wound contraction and healing	Fayazzadeh et al., 2014
Wound healing product	Sonochemical microemulsion process	GT+Alovera gel+almond oil	Wound healing with high growth and migration of fibroblast with antibacterial and antifungal property	Ghayempour et al., 2016a
Wound dressing patch.	Curcumin (Cur)-loaded poly (epsilon-caprolactone) (PCL)/ gum tragacanth (GT) scaffold membranes fabricated by electrospinning.	Curcumin	Provided the controlled release of curcumin for over 20 days for healing slow rate wounds	Ranjbar-Mohammadi et al., 2016.
Wound dressing nanofiber scaffolds	GT and poly (vinyl alcohol) (PVA) nanofibers were produced using electrospinning	GT	Wound dressing application, Antibacterial, scaffold for human fibroblast proliferation	Ranjbar-Mohammadi et al., 2013
Wound healing dressing hydrogel	poly (vinyl alcohol) (PVA), tragacanth gum (TG) and sodium alginate (SA) hydrogel entrapped moxifloxacin prepared by radiation method	Moxifloxacin	Drug released through non-Fickian mechanism with thrombogenicity (82.43+/-1.54%),	Singh et al., 2016

applications, owing to its biocompatible properties, slow, and sustained release of desirable therapeutics, porous matrix structure, biodegradability for prolonged administration of the drug in the target tissue. However, the very properties that make GT-nanocomposite based formulations an advanced desirable technology for therapeutics and drug invention may also lead to adverse health effects that demands proper safely evaluation of its pharmacodynamics and bio- interaction mechanisms before proceeding for a clinical trial.

3.10 CONCLUSION AND FUTURE PROSPECTIVE

GT is a natural polysaccharide based excipient resource with established pharmaceutical potentials. The present chapter distinguished recent advances in innovative pharmaceutical applications of biodegradable polymer GT, along with their possible formulation/modifications, in order to provide a compiled knowledge on its therapeutic applications in special reference to varied hydrogel-based tissue engineering scaffold materials, skin repair, nerve regeneration, bone repair, dental repair, nano-encapsulation, controlled site-specific drug delivery. Detailed scientific investigations on nano-particles and other forms of nano-system based GT formulations can be searched to produce an immense amount of very foretelling nanomedicine applicants. There have been raising attention in target specific, plant-based nano-therapeutics in pharmaceutical industries, largely expanding global market. Attributes of biocompatibility, non-toxicity, low cost, abundance, and potential applications of GT as an excipient in the areas of nanotherapeutic enhancement waited to add copious economic impact on global excipient market.

ACKNOWLEDGMENT

The authors would like to thank Professor Debasish Chowdhury, Institute of Advanced Study in Science and Technology (IASST, Guwahati, India), for his kind review and valuable editing suggestions for the improvement of the present review article.

KEYWORDS

- **biopolymer**
- **drug delivery**
- **excipient**
- **gum tragacanth**
- **nanocomposite**
- **polysaccharides**
- **tissue engineering**

REFERENCES

Abdolhossein, M. A. A. H., Homayon, A., & Iran, R., (2005). The effect of tragacanth mucilage on the healing of full-thickness wound in rabbit. *Arch Iranian Med.*, *8*(4), 257–262.

Abdolmaleki, K., Mohammadifar, M. A., Mohammadi, R., Fadavi, G., & Meybodi, N. M., (2016). The effect of pH and salt on the stability and physicochemical properties of oil-in-water emulsions prepared with gum tragacanth. *Carbohydr Polym*, *140*, 342–348.

Akhgari, A., Abbaspour, M. R., & Pirmoradi, S., (2011). Preparation and evaluation of pellets using acacia and tragacanth by extrusion-spheronization. *Daru*, *19*(6), 417–423.

Alchihab, M., Destain, J., Aguedo, M., Wathelet, J. P., & Thonart, P., (2010). The utilization of gum tragacanth to improve the growth of *Rhodotorula aurantiaca* and the production of gamma-decalactone in large scale. *Appl. Biochem. Biotechnol.*, *162*(1), 233–241.

Alvandimanesh, A., Sadrjavadi, K., Akbari, M., & Fattahi, A., (2017). Optimization of de-esterified tragacanth microcapsules by computational fluid dynamic and the Taguchi design with purpose of the cell encapsulation. *Int. J. Biol. Macromol.*, *105*(Part 1), 17–26.

Anderson, D. M., (1989). Evidence for the safety of gum tragacanth (*Asiatic astragalus* spp.) and modern criteria for the evaluation of food additives. *Food Additives and Contaminants*, *6*(1), 1–12.

Anderson, D. M., Ashby, P., Busuttil, A., Kempson, S. A., & Lawson, M. E., (1984). Transmission electron microscopy of heart and liver tissues from rats fed with gums Arabic and tragacanth. *Toxicology Letters*, *21*(1), 83–89.

Anderson, D. M., Busuttil, A., Kempson, S. A., & Penman, D. W., (1986). Transmission electron microscopy of jejunum, ileum, and caecum tissues from rats fed with gums Arabic, karaya and tragacanth. *Toxicology*, *41*(1), 75–82.

Anderson, D. M., Howlett, J. F., & McNab, C. G., (1985). The amino acid composition of the proteinaceous component of gum tragacanth (*Asiatic astragalus* spp.). *Food Additives and Contaminants*, *2*(4), 231–235.

Asghari-Varzaneh, E., Shahedi, M., & Shekarchizadeh, H., (2017). Iron microencapsulation in gum tragacanth using solvent evaporation method. *Int. J. Biol. Macromol.*, *103*, 640–647.

Azarikia, F., & Abbasi, S., (2010). On the stabilization mechanism of Doogh (Iranian yogurt drink) by gum tragacanth. *Food Hydrocolloids*, *24*(4), 358–363.

Bagheri, S. M., Keyhani, L., Heydari, M., & Dashti, R. M., (2015). Antinociceptive activity of Astragalus gummifer gum (gum tragacanth) through the adrenergic system: An *in vivo* study in mice. *J. Ayurveda Integr. Med.*, *6*(1), 19–23.

Banerjee, R., & Puniyani, R. R., (2000). Effects of clove oil-phospholipid mixtures on rheology of gum tragacanth - possible application for surfactant action on mucus gel simulants. *Biomed. Mater. Eng.*, *10*(3/4), 189–197.

Bhardwaj, T. R., Kanwar, M., Lal, R., & Gupta, A., (2000). Natural gums and modified natural gums as sustained-release carriers. *Drug Dev. Ind. Pharm.*, *26*(10), 1025–1038.

Brambilla, L., Riedo, C., Baraldi, C., Nevin, A., Gamberini, M. C., D'Andrea, C., Chiantore, O., Goidanich, S., & Toniolo, L., (2011). Characterization of fresh and aged natural ingredients used in historical ointments by molecular spectroscopic techniques: IR, Raman and fluorescence. *Analytical and Bioanalytical Chemistry*, *401*(6), 1827–1837.

Chenlo, F., Moreira, R., & Silva, C., (2010). Rheological behavior of aqueous systems of tragacanth and guar gums with storage time. *Journal of Food Engineering*, *96*(1), 107–113.

Comert, F., Azarikia, F., & Dubin, P. L., (2017). Polysaccharide zeta-potentials and protein-affinity. *Phys. Chem. Chem. Phys.*, *19*(31), 21090–21094.

Dehghan-Niri, M., Tavakol, M., Vasheghani-Farahani, E., & Ganji, F., (2015). Drug release from enzyme-mediated in situ-forming hydrogel based on gum tragacanth-tyramine conjugate. *J. Biomater. Appl.*, *29*(10), 1343–1350.

Eastwood, M. A., Brydon, W. G., & Anderson, D. M., (1984). The effects of dietary gum tragacanth in man. *Toxicology Letters*, *21*(1), 73–81.

FAO, (1992). *Tragacanth Gum.* (Food and nutrition paper 53) FAO, Rome.

FAO, (1995). *Gums, Resins and Latexes of Plant Origin.* FAO (Non-wood forest products 6) FAO, Rome.

Farahmandfar, R., Mohseni, M., & Asnaashari, M., (2017). Effects of quince seed, almond, and tragacanth gum coating on the banana slices properties during the process of hot air drying. *Food Sci. Nutr.*, *5*(6), 1057–1064.

Farzi, M., Emam-Djomeh, Z., & Mohammadifar, M. A., (2013). A comparative study on the emulsifying properties of various species of gum tragacanth. *Int. J. Biol. Macromol.*, *57*, 76–82.

Fattahi, A., Sadrjavadi, K., Golozar, M. A., Varshosaz, J., Fathi, M. H., & Mirmohammad-Sadeghi, H., (2013). Preparation and characterization of oligochitosan-tragacanth nanoparticles as a novel gene carrier. *Carbohydr. Polym.*, *97*(2), 277–283.

Fayazzadeh, E., Rahimpour, S., Ahmadi, S. M., Farzampour, S., Sotoudeh, A. M., Boroumand, M. A., & Ahmadi, S. H., (2014). Acceleration of skin wound healing with tragacanth (Astragalus) preparation: An experimental pilot study in rats. *Acta Med Iran*, *52*(1), 3–8.

Gavlighi, H. A., Meyer, A. S., Zaidel, D. N. A., Mohammadifar, M. A., & Mikkelsen, J. D., (2013). Stabilization of emulsions by gum tragacanth (Astragalus spp.) correlates to the galacturonic acid content and methoxylation degree of the gum. *Food Hydrocolloids*, *31*(1), 5–14.

Gavlighi, H. A., Michalak, M., Meyer, A. S., & Mikkelsen, J. D., (2013). Enzymatic depolymerization of gum tragacanth: Bifidogenic potential of low molecular weight oligosaccharides. *J. Agric. Food Chem.*, *61*(6), 1272–1278.

Ghayempour, S., & Montazer, M., (2017a). Tragacanth nanocapsules containing Chamomile extract prepared through sono-assisted W/O/W microemulsion and UV cured on cotton fabric. *Carbohydr. Polym.*, *170*, 234–240.

Ghayempour, S., & Montazer, M., (2017b). Ultrasound irradiation based in-situ synthesis of star-like Tragacanth gum/zinc oxide nanoparticles on cotton fabric. *Ultrason Sonochem*, *34*, 458–465.

Ghayempour, S., & Montazer, M., (2018). A modified microemulsion method for fabrication of hydrogel Tragacanth nanofibers. *Int. J. Biol. Macromol.*, *115*, 317–323.

Ghayempour, S., Montazer, M., & Mahmoudi, R. M., (2015). Tragacanth gum as a natural polymeric wall for producing antimicrobial nanocapsules loaded with plant extract. *Int. J. Biol. Macromol.*, *81*, 514–520.

Ghayempour, S., Montazer, M., & Mahmoudi, R. M., (2016a). Encapsulation of aloe vera extract into natural tragacanth gum as a novel green wound healing product. *Int. J. Biol. Macromol.*, *93*(Part A), 344–349.

Ghayempour, S., Montazer, M., & Mahmoudi, R. M., (2016b). Tragacanth gum biopolymer as reducing and stabilizing agent in biosonosynthesis of urchin-like ZnO nanorod arrays: A low cytotoxic photocatalyst with antibacterial and antifungal properties. *Carbohydr. Polym.*, *136*, 232–241.

Ghorbani, G. S., Ghorbani, G. E., Mohammadifar, M. A., & Zargaraan, A., (2014). Complexation of sodium caseinate with gum tragacanth: Effect of various species and rheology of coacervates. *Int. J. Biol. Macromol.*, *67*, 503–511.

Gupta, V. K., Sood, S., Agarwal, S., Saini, A. K., & Pathania, D., (2018). Antioxidant activity and controlled drug delivery potential of tragacanth gum-cl- poly (lactic acid-co-itaconic acid) hydrogel. *Int. J. Biol. Macromol.*, *107*(Part B), 2534–2543.

Gursoy, A., Karakus, D., & Okar, I., (1999). Polymers for sustained release formulations of dipyridamole-alginate microspheres and tabletted microspheres. *J. Microencapsul.*, *16*(4), 439–452.

Haeri, S. M., Sadeghi, Y., Salehi, M., Farahani, R. M., & Mohsen, N., (2016). Osteogenic differentiation of human adipose-derived mesenchymal stem cells on gum tragacanth hydrogel. *Biologicals*, *44*(3), 123–128.

Hagiwara, A., Boonyaphiphat, P., Kawabe, M., Naito, H., Shirai, T., & Ito, N., (1992). Lack of carcinogenicity of tragacanth gum in B6C3F1 mice. *Food Chem. Toxicol.*, *30*(8), 673–679.

Hagiwara, A., Tanaka, H., Tiwawech, D., Shirai, T., & Ito, N., (1991). Oral toxicity study of tragacanth gum in B6C3F1 mice: development of squamous-cell hyperplasia in the forestomach and its reversibility. *Journal of Toxicology and Environmental Health*, *34*(2), 207–218.

Hemmati, K., & Ghaemy, M., (2016). Synthesis of new thermo/pH-sensitive drug delivery systems based on tragacanth gum polysaccharide. *Int. J. Biol. Macromol.*, *87*, 415–425.

Hemmati, K., Masoumi, A., & Ghaemy, M., (2016). Tragacanth gum-based nanogel as a superparamagnetic molecularly imprinted polymer for quercetin recognition and controlled release. *Carbohydr. Polym.*, *136*, 630–640.

Iqbal, Z., Khan, R., Nasir, F., Khan, J. A., Rashid, A., Khan, A., & Khan, A., (2011). Preparation and *in-vitro in-vivo* evaluation of sustained release matrix diclofenac sodium tablets using PVP-K90 and natural gums. *Pak. J. Pharm. Sci.*, *24*(4), 435–443.

Jabri, T., Imran, M., Shafiullah, R. K., Ali, I., Arfan, M., & Shah, M. R., (2018). Fabrication of lecithin-gum tragacanth muco-adhesive hybrid nano-carrier system for in-vivo performance of Amphotericin B. *Carbohydr. Polym.*, *194*, 89–96.

Jain, A., Thakur, D., Ghoshal, G., Katare, O. P., & Shivhare, U. S., (2016). Characterization of microencapsulated beta-carotene formed by complex coacervation using casein and gum tragacanth. *Int. J. Biol. Macromol.*, *87*, 101–113.

Kedmi, S., & Katzenelson, E., (1978). A rapid quantitative fluorescent antibody assay of polioviruses using Tragacanth Gum. *Archives of Virology*, *56*(4), 337–340.

Keshtkaran, M., Mohammadifar, M. A., Asadi, G. H., Nejad, R. A., & Balaghi, S., (2013). Effect of gum tragacanth on rheological and physical properties of a flavored milk drink made with date syrup. *J. Dairy Sci.*, *96*(8), 4794–4803.

Kolangi, F., Memariani, Z., Bozorgi, M., Mozaffarpur, S. A., & Mirzapour, M., (2018). Herbs with potential nephrotoxic effects according to traditional Persian medicine: Review and assessment of scientific evidence. *Curr. Drug Metab.*, *19*(7), 628–637.

Komeilyfard, A., Fazel, M., Akhavan, H., & Mousakhani, G. A., (2017). Effect of Angum gum in combination with tragacanth gum on rheological and sensory properties of ketchup. *J. Texture Stud.*, *48*(2), 114–123.

Koppang, R., Berg, E., Dahm, S., Real, C., & Floystrand, F., (1995). A method for testing denture adhesives. *J. Prosthet. Dent.*, *73*(5), 486–491.

Koshani, R., & Aminlari, M., (2017). Physicochemical and functional properties of ultrasonic-treated tragacanth hydrogels cross-linked to lysozyme. *Int. J. Biol. Macromol.*, *103*, 948–956.

Kurt, A., Cengiz, A., & Kahyaoglu, T., (2016). The effect of gum tragacanth on the rheological properties of salep based ice cream mix. *Carbohydr. Polym.*, *143*, 116–123.

Liebert, M. A., (1987). Final report on the safety assessment of tragacanth gum. *Journal of the American College of Toxicology*, *6*(1), 1–22.

Lopez-Castejon, M. L., Bengoechea, C., Garcia-Morales, M., & Martinez, I., (2016). Influence of tragacanth gum in egg white based bioplastics: Thermomechanical and water uptake properties. *Carbohydr. Polym.*, *152*, 62–69.

Mallakpour, S., Abdolmaleki, A., & Tabesh, F., (2018). Ultrasonic-assisted manufacturing of new hydrogel nanocomposite biosorbent containing calcium carbonate nanoparticles and tragacanth gum for removal of heavy metal. *Ultrason. Sonochem.*, *41*, 572–581.

Masoumi, A., & Ghaemy, M., (2014). Removal of metal ions from water using nanohydrogel tragacanth gum-g-polyamidoxime: Isotherm and kinetic study. *Carbohydr. Polym.*, *108*, 206–215.

Mohamed, H. A., Radwan, R. R., Raafat, A. I., & Ali, A. E., (2018). Antifungal activity of oral (Tragacanth/acrylic acid) Amphotericin B carrier for systemic candidiasis: *In vitro* and *in vivo* study. *Drug Deliv. Transl. Res.*, *8*(1), 191–203.

Mohammadifar, M. A., Musavi, S. M., Kiumarsi, A., & Williams, P. A., (2006). Solution properties of tragacanthin (water-soluble part of gum tragacanth exudate from Astragalus gossypinus). *Int. J. Biol. Macromol.*, *38*(1), 31–39.

Montazer, M., Keshvari, A., & Kahali, P., (2016). Tragacanth gum/nanosilver hydrogel on cotton fabric: In-situ synthesis and antibacterial properties. *Carbohydr. Polym.*, *154*, 257–266.

Moradi, S., Taran, M., & Shahlaei, M., (2018). Investigation on human serum albumin and Gum Tragacanth interactions using experimental and computational methods. *Int. J. Biol. Macromol.*, *107*(Part B), 2525–2533.

Mostafavi, F. S., Kadkhodaee, R., Emadzadeh, B., & Koocheki, A., (2016). Preparation and characterization of tragacanth-locust bean gum edible blend films. *Carbohydr. Polym.*, *139*, 20–27.

Nasiri, M., Barzegar, M., Sahari, M. A., & Niakousari, M., (2018). Application of Tragacanth gum impregnated with Satureja khuzistanica essential oil as a natural coating for enhancement of postharvest quality and shelf life of button mushroom (Agaricus bisporus). *Int. J. Biol. Macromol.*, *106*, 218–226.

Niknia, N., & Kadkhodaee, R., (2017). Gum tragacanth-polyvinyl alcohol cryogel and xerogel blends for oral delivery of silymarin: Structural characterization and mucoadhesive property. *Carbohydr. Polym.*, *177*, 315–323.

Nur, M., & Vasiljevic, T., (2018). Insulin inclusion into a Tragacanth Hydrogel: An oral delivery system for insulin. *Materials (Basel)*, *11*(1), 79.

Pauk, V., Pluhacek, T., Havlicek, V., & Lemr, K., (2017). Ultra-high performance supercritical fluid chromatography-mass spectrometry procedure for analysis of monosaccharides from plant gum binders. *Anal. Chim. Acta.*, *989*, 112–120.

Rahmani, Z., Sahraei, R., & Ghaemy, M., (2018). Preparation of spherical porous hydrogel beads based on ion-crosslinked gum tragacanth and graphene oxide: Study of drug delivery behavior. *Carbohydr. Polym.*, *194*, 34–42.

Ranjbar-Mohammadi, M., & Bahrami, S. H., (2015). Development of nanofibrous scaffolds containing gum tragacanth/poly (epsilon-caprolactone) for application as skin scaffolds. *Mater. Sci. Eng. C. Mater. Biol. Appl.*, *48*, 71–79.

Ranjbar-Mohammadi, M., & Bahrami, S. H., (2016). Electrospun curcumin loaded poly(epsilon-caprolactone)/gum tragacanth nanofibers for biomedical application. *Int. J. Biol. Macromol.*, *84*, 448–456.

Ranjbar-Mohammadi, M., (2018). Production of cotton fabrics with durable antibacterial property by using gum tragacanth and silver. *Int. J. Biol. Macromol.*, *109*, 476–482.

Ranjbar-Mohammadi, M., Bahrami, S. H., & Joghataei, M. T., (2013). Fabrication of novel nanofiber scaffolds from gum tragacanth/poly(vinyl alcohol) for wound dressing application: *In vitro* evaluation and antibacterial properties. *Mater. Sci. Eng. C. Mater. Biol. Appl.*, *33*(8), 4935–4943.

Ranjbar-Mohammadi, M., Prabhakaran, M. P., Bahrami, S. H., & Ramakrishna, S., (2016). Gum tragacanth/poly(l-lactic acid) nanofibrous scaffolds for application in regeneration of peripheral nerve damage. *Carbohydr. Polym.*, *140*, 104–112.

Ranjbar-Mohammadi, M., Prabhakaran, M. P., Hajir, B. S., & Ramakrishna, S., (2017). Corrigendum to "Gum tragacanth/poly (l-lactic acid) nanofibrous scaffolds for application in regeneration of peripheral nerve damage" (*Carbohydr. Polym. J.*, *140*, 104–112). *Carbohydr. Polym.*, *160*, 212.

Ranjbar-Mohammadi, M., Rabbani, S., Bahrami, S. H., Joghataei, M. T., & Moayer, F., (2016). Antibacterial performance and in vivo diabetic wound healing of curcumin loaded gum tragacanth/poly(epsilon-caprolactone) electrospun nanofibers. *Mater. Sci. Eng. C. Mater. Biol. Appl.*, *69*, 1183–1191.

Ranjbar-Mohammadi, M., Zamani, M., Prabhakaran, M. P., Bahrami, S. H., & Ramakrishna, S., (2016). Electrospinning of PLGA/gum tragacanth nanofibers containing tetracycline hydrochloride for periodontal regeneration. *Mater. Sci. Eng. C. Mater. Biol. Appl.*, *58*, 521–531.

Rao, K., Imran, M., Jabri, T., Ali, I., Perveen, S., Shafiullah, A. S., & Shah, M. R., (2017). Gum tragacanth stabilized green gold nanoparticles as cargos for Naringin loading: A morphological investigation through AFM. *Carbohydr. Polym.*, *174*, 243–252.

Rasul, A., Iqbal, M., Murtaza, G., Waqas, M. K., Hanif, M., Khan, S. A., & Bhatti, N. S., (2010). Design, development and *in-vitro* evaluation of metoprolol tartrate tablets containing xanthan-tragacanth. *Acta. Pol. Pharm.*, *67*(5), 517–522.

Sadat, H. M., Hemmati, K., & Ghaemy, M., (2016). Synthesis of nanohydrogels based on tragacanth gum biopolymer and investigation of swelling and drug delivery. *Int. J. Biol. Macromol.*, *82*, 806–815.

Sadeghi, S., Rad, F. A., & Moghaddam, A. Z., (2014). A highly selective sorbent for removal of Cr(VI) from aqueous solutions based on Fe(3)O(4)/poly(methyl methacrylate) grafted Tragacanth gum nanocomposite: Optimization by experimental design. *Mater. Sci. Eng. C. Mater. Biol. Appl.*, *45*, 136–145.

Sadrjavadi, K., Shahbazi, B., & Fattahi, A., (2018). De-esterified tragacanth-chitosan nano-hydrogel for methotrexate delivery, optimization of the formulation by Taguchi design. *Artif. Cells Nanomed. Biotechnol.*, 1–11.

Sahraei, R., & Ghaemy, M., (2017). Synthesis of modified gum tragacanth/graphene oxide composite hydrogel for heavy metal ions removal and preparation of silver nanocomposite for antibacterial activity. *Carbohydr. Polym.*, *157*, 823–833.

Salamanca, C. H., Yarce, C. J., Moreno, R. A., Prieto, V., & Recalde, J., (2018). Natural gum-type biopolymers as potential modified nonpolar drug release systems. *Carbohydr. Polym.*, *189*, 31–38.

Shahiwala, A., & Zarar, A., (2018). Pharmaceutical product lead optimization for better *in vivo* bioequivalence performance: A case study of diclofenac sodium extended-release matrix tablets. *Curr. Drug. Deliv.*, *15*(5), 705–715.

Sharma, H., Sharma, B. D., Talukder, S., & Ramasamy, G., (2015). Utilization of gum tragacanth as bind enhancing agent in extended restructured mutton chops. *J. Food Sci. Technol.*, *52*(3), 1626–1633.

Shimotoyodome, A., Kobayashi, H., Nakamura, J., Tokimitsu, I., Hase, T., Inoue, T., Matsukubo, T., & Takaesu, Y., (2006). Reduction of saliva-promoted adhesion of Streptococcus mutans MT8148 and dental biofilm development by tragacanth gum and yeast-derived phosphomannan. *Biofouling*, *22*(3/4), 261–268.

Shiroodi, S. G., Mohammadifar, M. A., Gorji, E. G., Ezzatpanah, H., & Zohouri, N., (2012). Influence of gum tragacanth on the physicochemical and rheological properties of kashk. *J. Dairy Res.*, *79*(1), 93–101.

Singh, B., & Sharma, V., (2014). Influence of polymer network parameters of tragacanth gum-based pH-responsive hydrogels on drug delivery. *Carbohydr. Polym.*, *101*, 928–940.

Singh, B., & Sharma, V., (2017). Crosslinking of poly(vinylpyrrolidone)/acrylic acid with tragacanth gum for hydrogels formation for use in drug delivery applications. *Carbohydr. Polym.*, *157*, 185–195.

Singh, B., Varshney, L., & Francis, S., (2016). Rajneesh, Designing tragacanth gum based sterile hydrogel by radiation method for use in drug delivery and wound dressing applications. *Int. J. Biol. Macromol.*, *88*, 586–602.

Smee, D. F., Sidwell, R. W., Huffman, J. H., Huggins, J. W., Kende, M., & Verbiscar, A. J., (1996). Antiviral activities of tragacanthin polysaccharides on Punta Toro virus infections in mice. *Chemotherapy*, *42*(4), 286–293.

Strobel, S., Ferguson, A., & Anderson, D. M., (1986). Immunogenicity, immunological cross-reactivity and non-specific irritant properties of the exudate gums, Arabic, karaya and tragacanth. *Food Additives and Contaminants*, *3*(1), 47–56.

Tavakol, M., Dehshiri, S., & Vasheghani-Farahani, E., (2016). Electron beam irradiation crosslinked hydrogels based on tyramine conjugated gum tragacanth. *Carbohydr. Polym.*, *152*, 504–509.

Tavakol, M., Vasheghani-Farahani, E., Mohammadifar, M. A., Soleimani, M., & Hashemi-Najafabadi, S., (2016). Synthesis and characterization of an *in situ* forming hydrogel using tyramine conjugated high methoxyl gum tragacanth. *J. Biomater. Appl.*, *30*(7), 1016–1025.

Tonyali, B., Cikrikci, S., & Oztop, M. H., (2018). Physicochemical and microstructural characterization of gum tragacanth added whey protein-based films. *Food Res. Int.*, *105*, 1–9.

Torres, M. D., Moreira, R., Chenlo, F., & Vazquez, M. J., (2012). Water adsorption isotherms of carboxymethyl cellulose, guar, locust bean, tragacanth and xanthan gums. *Carbohydr. Polym.*, *89*(2), 592–598.

Verbeken, D., Dierckx, S., & Dewettinck, K., (2003). Exudate gums: Occurrence, production, and applications. *Appl. Microbiol. Biotechnol.*, *63*(1), 10–21.

Zarekhalili, Z., Bahrami, S. H., Ranjbar-Mohammadi, M., & Milan, P. B., (2017). Fabrication and characterization of PVA/Gum tragacanth/PCL hybrid nanofibrous scaffolds for skin substitutes. *Int. J. Biol. Macromol.*, *94*(Part A), 679–690.

CHAPTER 4

Application Potential of Pectin in Drug Delivery

PRITISH KUMAR PANDA,[1] GAYTRI GOUR,[1] SHIVANI SARAF,[1] ANKITA TIWARI,[1] AMIT VERMA,[1] ANKIT JAIN,[2] and SANJAY K. JAIN[1]

[1]*Pharmaceutics Research Projects Laboratory, Department of Pharmaceutical Sciences, Dr. Hari Singh Gour Central University, Sagar (M.P.) – 470003, India*

[2]*Institute of Pharmaceutical Research, GLA University, NH-2, Mathura-Delhi Road, Mathura (U.P.) – 281406, India*

ABSTRACT

Pectin, a naturally occurring, biodegradable heteropolysaccharides, comprised of (1,4)-linked α-d-galacturonic acid entities and numerous neutral sugars like D-galactose, L-arabinose, and D-xylose, etc., has grabbed the attention of scientists in last few years because of its biodegradability, biocompatibility, and non-toxicity. It has found promising applications in an arena of drug delivery and biomedical field. On the basis of its degree of esterification (DE), it has been classified into two major groups, i.e., low methoxy pectin (LMP) and high methoxyl pectin (HMP). The pectin chains combine to form the three-dimensional frameworks leading to the formation of a gel. The pectin properties can be enhanced and chemically altered to form composites, blends for a multiplicity of pharmaceutical applications. It has been utilized potentially as a polymer for targeted drug delivery. This chapter deals with the application potential of pectin in the development of drug delivery systems.

4.1 INTRODUCTION

Pectin is a biodegradable heterogeneous polysaccharide, which occurs in the plant's cell wall and is also a principal constituent of the middle lamella where it keeps cells close to each other or conterminous network of pectin provides the cell wall with the capability to resist compression. Pectin composition in the plant cell wall is shown in Figure 4.1 (Wang et al., 2014; Zhang et al., 2015). Chemically, it is a complex polysaccharide comprised of 1,4-linked α-d-galactosyluronic acid (GalpA) units (Figure 4.2). There are mainly three pectin polysaccharides like homogalacturonan (HG), rhamnogalacturonan-I, and rhamnogalacturonan-II (Figure 4.3) (Chung et al., 2017). Approximately 65% of pectin is found in HG, and it is partially methylated at C-6 and O-acetylated at 2nd or 3rd position. While rhamnogalacturonan-I contains pectin in the range of 20 to 35% and having more complexity in structure as compared to HG. Chemically it contains about 100 repeating units of (1,2)-α-L-Rha-(1,4)-α-D GalpA. Neutral side chain substitution at O-4 on rhamnose moiety is greater, and it also includes arabinogalactan I and II, arabinan, and galactan. Pectin composition in rhamnogalacturonan II is approximately 10%, and structurally it is more complex. It is present in minor quantity and plays a significant role in the formation of the plant cell wall (Noreen et al., 2017)

Pectin is highly fibrous in nature which is helpful to reduce cholesterol and serum glucose as well as having anti-cancer properties and is also capable of triggering apoptosis in adenocarcinoma cells (Mishra et al., 2012). Currently, China is the prime hub for the citrus plantation, and approximately 70% of canned citrus is collected by China segments globally. Citrus peel is the chief source for obtaining pectin (up to 20–30% dry weight), which is a huge waste material in the fruit juice industry after extraction. Grapefruit peel contains more rhamnogalacturonan-1among all pectin as compared with other citrus fruits. One of the easiest and economical methods for the extraction of pectin is the treatment of raw materials with hot water in the presence of acid for a long duration. Since this method is quite a time consuming, therefore, another method has been discovered in which water is used as a dissolving medium and the energy is supplied into it through microwave for the extraction, where microwave- mediated extraction of pectin occurs from mango peel waste by hydrothermal means Other methods such as acoustic cavitation or ultrasound assisted extraction are economical methods used for the extraction of pectin (Wang et al., 2017). Pectin can be extracted from the different sources by various methods, and a general extraction process from the beet pulp has been outlined by a flow diagram in Figure 4.4.

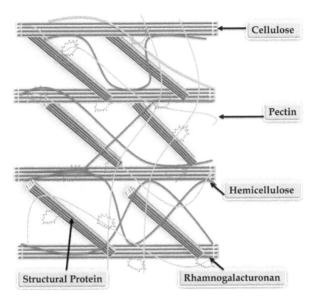

FIGURE 4.1 **(See color insert.)** Composition of pectin in the plant cell wall.

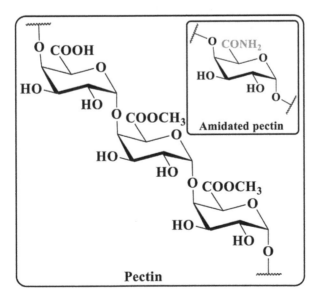

FIGURE 4.2 **(See color insert.)** Chemical structures of pectin and amidated pectin.

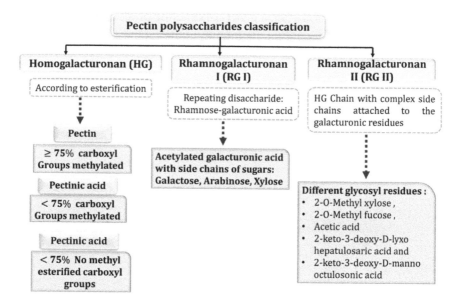

FIGURE 4.3   (See color insert.) Classification of pectin.

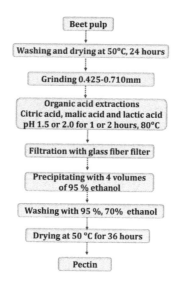

FIGURE 4.4   Flow diagram of the extraction process of pectin from beet pulp.

On the basis of processing methods, it has been divided as natural and modified pectin of high and low molecular mass (Mm), respectively

(Zhang et al., 2015). Moreover, HMP and LMP is another way of classification based on DE. HMP having more than 50% DE also requires soluble solid more than 55% at low pH with a range of 2.8 to 3.4 for gel formation. LMP having less than 50% DE forms a gel with the high calcium ion (Ca^{++} ions) concentration over a pH range of 2.5–6.5. Carboxyl groups present in pectin are converted to amide form approximately 15–25% after treating with the ammonia and this amidated pectin is sensible to Ca^{++} ions. Pectin is more stable at acidic pH with a range of 2–4.5. HMP requires lowering of gelling temperature, which may be obtained by the de-esterification at pH < 2 while the increase in gelling temperature is required for medium esterified pectin. HMP degrades at pH from 4.5 to 10, which is faster than LMP. It is due to the lack of methyl ester group, pectic acid, and pectates, which are stable at this pH range (Tiwary and Rana, 2016; Thibault and Ralet, 2003). Pectin has good solubility in pure water; similarly, its monovalent cationic salts of pectinic and pectic acids also depict solubility in the water while the di- and trivalent cationic salts show minute solubility or insolubility. When dry powder of pectin comes in contact with water, it has a propensity to trap water very rapidly and form clumps. Various factors which may affect the gelling property of pectin are molecular weight, pH, temperature, DE, the concentration of the preparation, and presence of counter ions in solution, etc. (Sriamornsak, 2003). Greater pH and very low concentration of soluble solids are favorable for the gelling process while lower pH and high concentration of soluble solids need greater gelling temperature, which causes fragmented gelation. The gelling temperature may drop with a decrease in DE (about 58–62%). De-esterification of HMP yields distinctive charge distribution, which increases gel formation with calcium ions by ionotropic interaction process. This charge modified HMP shows the potential to strongly bind with the cationic proteins. The presence of calcium ion is enhanced with the increase in ionic strength, decrease of soluble solids, and an increase in pH. Moreover, the molecular weight, charge density, charge distribution and the degree of acetylation, are found to be key players in determining the gelling properties of pectin and also its DE properties (Tiwary and Rana, 2016; May, 1990). Various sources (Table 4.1) of pectin such as apple pomace and citrus fruits peel (lemon, orange, lime, etc.) are the most common sources of pectin. Based on dry matter approximately 20–30% of pectin is present in citrus peel, whereas about 10–15% is present in apple pomace (Yadav et al., 2017). Pectin is widely used for jellies and jams besides, and it is also used for fruit products in the food industry, desserts, soft drinks, dairy products and in pharmaceuticals as an emulsifying, stabilizing, thickening, and gelling agent. Moreover, it is able to protect against numerous types of lifestyle modifying diseases, such as obesity, diabetes, constipation, and gall stone and various cancers including liver cancer, and colon cancer, etc. (Jiang et al., 2012).

TABLE 4.1   Various Sources of Pectin

High Pectin Fruit	Moderate Pectin Fruit	Low Pectin Fruit
Apple, crisp, and tart	Apples, ripe	Apricots
Blackberries, sour	Blackberries, ripe	Blueberries
Crabapples	Cherries, sour	Cherries, sweet
Cranberries	Chokecherries	Figs
Gooseberries	Elderberries	Grapes (Western Concord)
Lemons	Grape Juice, bottled	Peaches
Loganberries	Oranges	Pears
Quinces	Plums (not Italian)	Plums (Italian)
		Raspberries
		Strawberries

Pectin also has distinctive properties, which empower it to make matrices for the drug/protein entrapment and/or delivery into the cells. It finds extensive applications as a carrier for intestine/colon targeting like matrix tablets, gel, beads, and film-coated formulations, etc. (Sriamornsak, 2003). Mucoadhesive property of pectin is crucial for effective drug delivery to the particular site of the body. There could be two mechanisms behind this, and one may be the H bond formation with mucin due to the presence of –COOH group in pectin and second could be the electrostatic forces existing between pectin and mucin. This is responsible for the aggregate formation when it adheres with mucin, and it increases with an increasing amount of pectin. Both mucin and pectin have a negative charge, causing repulsion and unwinding of the polymer chain, which increases the adhesion (Chatterjee et al., 2017). Humans have almost a fixed microflora composition. Pectin is unchanged in the upper GI tract and decomposed by microflora present in the colon. Therefore, pectin-based drug delivery systems possess great potential for targeting colon-associated diseases. Nowadays, pectin derived matrices are used for controlled drug delivery (Liu et al., 2003). Natural pectin composed of dietary fibers is advised for the treatment of colon cancer. Its bioavailability and bioactivity have been enhanced by the modification to get modified pectin fragments having low molecular weight. It is also having low DE and investigated for anticancer potentials like restraining metastasis, induction of apoptosis, and regulation of immunological responses. This modified pectin is involved in recognition of ligand by galectin-3 (Do Prado et al., 2017). Moreover, antitumor activity is due to the involvement of probiotic activity, and the regulation of transformation-related micro RNA/oncogenes affects the cells of the colon and their immune response. Apple pectin is good for the treatment of colon cancer as it reduces fecal bacterial enzyme activity, scavenges free radicals, decreases DNA adducts through the regulation of microRNAs. Likewise, the antitumor activity can be increased through the chemical modification (Almeida et al., 2015), radiation, heating, and/or enzymes for pectin degradation to reduce its DE, e.g., the growth of HT-29 cells was suppressed by heated ginseng pectin (Cheng et al., 2011). Similarly, pectin with 20kGy of γ-irradiation curbed HT-29 cancer cells. The same report has been found with the pH-modified citrus pectin causes inhibition of tumor progression, metastases, and angiogenesis and induces apoptosis of prostate cancer cells. This modified pectin reduces colon cancer through inhibition of inflammation (Glinsky and Raz, 2009). Mucin-2 is a glycoprotein having O-glycan and is a Gal-3 ligand defectively expressed in colon cancer. Gal-3 up-regulates mucin-2 at the transcriptional level through the activation of AP-1, a type of transcription factor (Zhang et al., 2015). Therefore, drug delivery systems that target mucin-2 pathway may be a potential strategy for the treatment of colon cancer. Moreover, it has been described that the pectin has a good ability to increase the transit time through bile acid excretion, generation of fatty acids, fecal bulk, etc. As it is a soluble fiber, it goes through 100% fermentation in the colon and generates short-chain fatty acids (Lupton, 2000; Moore et al., 1998). These reduce colonic pH, which is supportive for the protection against colon cancer (Ferguson and Harris, 1996). Previous studies show the difference between fecal pH of a colon cancer patient and a normal person's fecal pH, which is 7 and 6.5, respectively. Here, 7-α-Dehydroxylase converts primary bile acid into secondary bile acid-like lithocholic acid and deoxycholic acid, which eventually cause colon cancer. Fermentation of pectin helps to reduce pH and cures colon cancer by inhibiting enzyme 7-α-Dehydroxylase at pH below 6.5 as well as decreases the solubility of free bile acids. These short-chain fatty acids also reduce colon hyper-proliferation by decreasing the expression of proto-oncogenes c-Fos and c-Jun. Citrus pectin, LMP, and HMP have been accounted earlier which increase the risk of colon cancer while apple pectin and citrus pectin reported lately to reduce the chances of colon cancer (Bauer et al., 1981). The biological function of pectin for curing colon cancer acts as an inhibitor of galectin-3; it also enhances the apoptosis activity of HT29 cells (Olano-Martin et al., 2002). Here, Galectin-3 binds with galactoside present in the surface of the cell, cytoplasm, nucleus, and also secreted itself by the cancer cells (Byrd and Bresalier, 2004). Pectin can have the potential to suppress colon cancer significantly when taken as a dietary supplement and exerts an anti-proliferative effect in mouse distal colon (Heitman et al., 1992; Umar et al., 2003).

4.2   CONVENTIONAL SYSTEMS

Since many years, pectin has been employed in numerous formulations like matrix tablets and film coatings, etc. Calcium pectinate-based carrier system has been reported for colon-targeted delivery by compression into tablets (Rubinstein et al., 1993). The potential of high methoxyl (HM)-pectin and combinations of calcium salts and low-methoxyl (LM) pectin have been investigated to prepare matrix tablets for administering different drugs for colon targeting. Ashford et al., (1993) depicted that HM-pectin, when utilized as a compression coat, evidenced the ability to safeguard the tablet mimicking transit from oral cavity to colon (Ashford et al., 1993). *In vivo* γ-scintigraphy affirmed that the tablets coated with pectin undergo disintegration in the colon. The *in-vitro* and *in-vivo* assessment of calcium pectinate for colon-specific preparations has been examined by Rubinstein and Radai (1995). In *in-vitro* studies, the varying conditions in the GI tract were mimicked, and the release of indomethacin (IND) from matrix tablets of calcium pectinate and compression-coated tablets with another layer of calcium pectinate was compared. The *in vivo* behavior of matrix tablets of calcium pectinate and that of compression-coated tablets bearing insulin was compared, succeeding the oral delivery in dogs. Compressed matrices could retain the drug in simulated gastrointestinal (GI) fluids before their break down by the action of various pectinolytic enzymes. Thus, these formulations could be employed for the colonic delivery of hydrophobic drugs. The extended insulin absorption is associated with the degradation of the drug carrier system in the large intestine of dogs. Uncoated tablets of calcium pectinate could not preclude insulin diffusion and commenced their release instantly after administration. Therefore, an extra protective coat might be necessitated for hydrophilic drugs. Ugurlu et al., (2007) revealed that pectin alone is inadequate to safeguard the drug loaded in the core tablets for delivering to the colon due to the degradation of pectin (Ugurlu et al., 2007). The inclusion of HPMC (5% w/w) with pectin in compression coat rendered a two hours lag time for drug release. The *in vivo* study revealed that, in all the healthy subjects, the drug was released from tablets in the colon (Hodges et al., 2009). Matrix tablets containing ropivacaine were prepared from pectin for colon-specific delivery. The effect of amidated pectin (Am.P) and a calcium salt of pectin (Ca.P), the inclusion of ethyl cellulose as an additive were inquired in a D-optimal mixture study. Even though neither of the two pectin forms can restrict the drug release to the required level; Am.P was observed to be ameliorating colonic delivery due to its higher susceptibility to degradation by the enzymes (Ahrabi et al., 2000).

Pectin-hydroxypropyl methylcellulose (HPMC) coated 5-aminosalicylic acid (5-ASA) tablets using different percentages of the above polymers were prepared. The *in vitro* outcomes depicted that 20% HPMC and 80% pectin exhibited better results and efficiently administered 5-ASA to the colon. They could be employed to develop an optimum system for colon-specific delivery (Turkoglu and Ugurlu, 2002). Matrix tablets of pectin were formulated by direct compression. The swelling and erosion properties were studied in different media. The pH of the medium affected the release of drug from the tablets. The mechanism of drug release and its release kinetics were influenced by the swelling and erosion characteristics. Thus, by regulating these factors, sustained release tablets could be developed easily (Sriamornsak et al., 2007).

4.3   RECENT PECTIN-BASED TARGETING SYSTEMS

4.3.1   HYDROGEL SYSTEM

Pectin-based hydrogels are of great attention in the field of Pharmaceutical industries, having the potential to absorb fluid within their polymeric networks. Hence, their biocompatible nature and good water absorbing property made them suitable for biomedical technology, organ regeneration, tissue engineering and drug delivery (Demir and Kerimoglu, 2015; Hong et al., 2014; Kim and Peppas, 2003; Morita et al., 2000). Table 4.2

indicates the various pectin based carrier systems with their diverse applications. Moreover, specifically, their swelling behavior is one of the significant parameters, which affects the mechanical strength, surface wettability, and diffusion behavior (Chen et al., 1999). Pectin-based hydrogel carriers are being fabricated to encourage their utility in colon targeting. After oral administration, it is difficult for the hydrogel system to pass via the upper parts of the GI tract and get delivered to the target site. This problematic condition can be overcome with the help of a pectin-based hydrogel system by providing stronger cross-linking interactions. It can be performed by the use of both a divalent basic calcium ion (Ca^{2+}) and oligochitosan, which has improved the stability of the hydrogel systems. In addition to this, these two cross-linkers also contribute to the improvement of the pectin-based hydrogel systems in colon targeting (Zhang et al., 2016). The pectin-based hydrogel can be opted as an alternative to oral or parenteral system owing to its vast benefits by developing microneedle, gel, etc. It includes advantages such as escaping of GI degradation, abstinence from the first pass hepatic metabolic effect, increased bioavailability, and improved patient compliance. This transformation to clinics is only promising by addressing the parameters which are utmost necessary to be considered like their manufacturing approach, regulatory requirements, safety/biocompatibility, adverse/immunological effects, disposability, long term safety profile, and frequent utilities. To provide abundant benefits of this polymer-based hydrogel system is proficient for carrying the pharmacological potent drugs, therapeutically narrow indexed drugs, and cosmetic ingredients (Eltayib et al., 2016; Chen et al., 2016; Park et al., 2015). Poly (methyl vinyl ether-comaleic anhydride) (PMVE/MA) is a synthetic copolymer of methyl vinyl ether and maleic anhydride. PMVE/MA hydrogels have the tendency to bind with a variety of plasticizers. Polyethylene glycol (PEG) is a commonly employed plasticizer for the formulation of PMVE/MA hydrogels. It was shown that the polyhydric alcohol-type plasticizers could also be cross-linked with PMVE/MA and manufactured by either heating or incubation at 25°C. A tripropylene glycol methyl ether (TPGME) plasticizer has been similarly be used to develop PMVE/MA cross-linked hydrogels (Demir et al., 2017). The dearth of satisfactory mechanical features limits the use of pectin hydrogel systems in biomedicine and specifically in tissue repair technology. Recently, double network hydrogels (DNH) have gained high interest and demand, which are made up of two contrasting networks. Recently a new approach like a pectin-Fe/polyacrylamide hybrid

DNH has been established by a simple multistep method. Here the Fe ions are kept together into a network consisted of pectin to generate strong ionic complexation. It is reversible and has the competency to withstand the mechanical strength. Apart from this, it also offers good toughness tensile strength, stiffness, and water absorption capacity. This type of new hydrogel system has excellent ability to widen the utility in tissue repair industries (Niu et al., 2017). A novel pH-sensitive gelatin/pectin hydrogels were developed by using various polymeric ratios. The polymers type and its degree of crosslinking have a great influence on swelling and release characteristics. The gelatin/pectin-based hydrogels were prepared with the help of a crosslinking agent, i.e., glutaraldehyde (GA). A number of parameters like an average molecular weight between crosslinking agents, swelling behavior, sol-gel transformation, volume fraction of polymer, porosity, and diffusion coefficient were evaluated. The hydrogels exhibit a high water uptake tendency and an enhanced release rate at low pH. The thermal stability was determined by both the DSC and TGA studies, while SEM result confirmed the porosity of the formulation. The formulation displayed non-Fickian diffusion. Characterization techniques revealed the successful formation of copolymer hydrogels. All these characteristics, like the pH-responsiveness, swelling ability, and drug release behavior, propose their probable use for site-specific drug delivery (Naeem et al., 2017). A highly potential new hydrogel networking system was recently developed which was prepared from pectin, a crosslinker of phosphate like bis [2-methacryloyloxy] and ethyl phosphate (BMEP) poly ((2- dimethyl-amino)ethyl methacrylate)) and exhibited dual response. These hydrogels systems have arranged by free radical polymerization method and are successfully used for encapsulation of 5-fluorouracil (5-FU), which is an anticancer drug. It has been utilized for the construction of silver nanoparticles. The networks of hydrogel and silver nanoparticles were assessed by different modern techniques like SEM, XRD, and DSC, etc. The 5-FU loaded hydrogels systems were investigated for *in vitro* release studies in the stomach as well as in intestine with different pH (pH 1.2 & pH 7.4) and temperature (25°C and 37°C). In addition to this, the antibacterial activity of the dual response hydrogel system based silver nanocomposites were tested by using zone of inhibition test against four different species of bacteria like *Bacillus cereus* (+ve), *Escherichia coli* (–ve), *Klebsiella pneumoniae* (–ve), and *Staphylococcus aureus* (+ve) (Eswaramma et al., 2017). An advanced biomimetic injectable hydrogel system has been established by incorporating hyaluronic acid-adipic dihydrazide and oligopeptide G RGDS (glycine–arginine–glycine–aspartate–serine)-grafted oxidized pectin. The hydrazide and aldehyde-derivatives were crosslinked with each other by the help of a covalent bond. The crosslinking made the system simple, non -toxic, feasible, less invasive, and translatable for easy regenerative applications in tissue engineering. Additionally, the chondrocyte behavior was dependent on pectin/hyaluronic acid composition and specifically the incorporation of G RGDS oligopeptide. The incorporation of G RGDS oligopeptide into the pectin/hyaluronic acid-based hydrogel systems could be competent to afford an active natural microenvironment, which helped the utility of chondrocyte and expedited chondrogenesis. Overall, the new novel injectable biomimetic hydrogel system was envisaged as an excellent tool for utility in the biomedical field for cartilage tissue regeneration (Chen et al., 2017). Novel food-grade pectin and starch-based hydrogel particles were prepared for probiotic action to the colon. *Lactobacillus plantarum* encapsulation in pectin/starch hydrogels was done by the extrusion method. Four batches with various concentrations ratio of pectin/starch were formulated, and in the hydrogel network, the distribution of *L. plantarum* was detected by microscopy as well as the viability of encapsulated cells in the hydrogels were detected against the varying conditions of the gastrointestinal tract (GIT) and salt solution of bile. Hence, it can be proposed that the pectin/starch based hydrogel systems can be considered as a transporter for oral use of the probiotic bacterial to the colon (Dafe et al., 2017). Pectin-based hydrogel films have been manufactured with immense potential in the biomedical industry and are also applicable in wound dressings. Simvastatin loaded cross-linked alginate-pectin hydrogel films have been developed by ionic crosslinking to boost the drug release behavior, mechanical properties, and the wound healing property. Alginate-pectin hydrocolloid films were prepared by crosslinking with calcium chloride ($CaCl_2$) in different concentrations (0.5–3% w/v) for 10 to 20 minutes because both $CaCl_2$ concentration and the contact time have an impact on the degree of cross-linking. The 0.5%–1% concentration of $CaCl_2$ for 2 min was suitable for effective crosslinking. Then the hydrogel system was characterized for their different properties such as physical, morphological, mechanical, thermal, *in vitro* drug release kinetics and the cytocompatibility studies. The *in vitro* cytotoxicity assay revealed that the system was biocompatible and non-toxic (Rezvanian et al., 2017). Tyramine-Pectin hydrogel system was arranged by cross-linking calcium with carboxyl moiety and phenol groups with peroxidase. Then, modification in pectin was done with sodium periodate at different molar ratios (2.5 to 20 mol%) by oxidation process and followed by the reductive amination with tyramine and sodium cyanoborohydride. The hydrogel system was tried as carriers for soybean hull peroxidase immobilization within the microbeads (Prokopijevic et al., 2017). A new nanocomposite hydrogel was developed by the combination of chitosan (CS), pectin, and montmorillonite clay. Different concentration of clay (0.5 and 2%), as well as polymer ratios (1:1, 1:2, and 2:1), was taken. The incorporation of clay into the hydrogel provides higher thermal stability along with good compression resistance (da Costa et al., 2016). Some recently developed pectin-based hydrogel systems have been intended for oral administration also permit the addition of pluronic, tween, sodium lauryl sulfate, etc., which might contribute to fit the release kinetics of the drug. The combination of different surfactants formed pectin-based hydrogels may become more competent in recent coming years (Marras-Marquez et al., 2015).

4.3.2 MICROSPHERES SYSTEM

Pectin microspheres containing 5-fluorouracil (5-FU) were designed using emulsion dehydration method and coating with Eudragit for colonic delivery. Several ratios of pectin and 5-FU (3:1 to 6:1) at different concentration ranges of emulsifiers (0.75%–1.5% wt/vol) were used with suitable stirring speed. Oil-in-oil solvent evaporation technique was commonly employed for eudragit-coating by using a definite ratio of coat and core (5:1). The evaluation of formulated pectin based microspheres is performed via several parameters, i.e., surface morphology, size, and size distribution of particle, percentage drug entrapment, swelling potential and *in vitro* release studies of the drug in simulated GI fluids. In addition to it, the analysis was conducted in simulated colonic fluid (SCF) in rat with 2% caecal matter Biodistribution study was also done in the colon of albino species of rats. Eudragit coated pectin-based microspheres were analyzed with two parameters, such as pharmacokinetic analysis and release characteristics of FU, which was pH dependent. Moreover, it was also observed that at acidic pH, the drug was gradually released from the system while in basic medium (pH 7.4) the release was observed to be faster (Paharia et al., 2007).

Polymeric microspheres, i.e., pectin-based microspheres were prepared by spray-drying method and loaded with ondansetron hydrochloride for the delivery via the nasal route. They escaped from the first-pass metabolism in the liver and improved the residence time. Thirty-two full experimental factorial designs were employed for the optimization of the developed formulation. In addition to this, stability study, zeta potential study, drug release kinetics, *ex vivo* penetration study, histological examination, and *in vivo* evaluation and differential scanning calorimetry (DSC), scanning electron microscopy (SEM) and X-ray diffraction study (XRD) were also employed for characterization of the microsphere system (Mahajan et al., 2012). Pectin–metronidazole (PT–ME) prodrug containing microspheres were synthesized by emulsion-dehydration technique, and it was confirmed by various spectroscopic studies. Microspheres exhibited complete release of drug in SGF, and the *in-vivo* studies were performed, which efficiently delivered the drug to the colon site (Vaidya et al., 2015). Pectin-based microspheres containing aceclofenac were formulated for chronopharma-cological delivery for rheumatoid arthritis. Solvent evaporation technique was used for coating Eudragit S 100 on microspheres. The drug release was found to improve (8 hours) in the presence of cecal contents. The *in* vivo results confirmed the application of the developed system for the management of RA (Ramasamy et al., 2013).

4.3.3 NANOPARTICLES SYSTEM

Recently, the pectin-conjugated systems were used in the treatment of cancer. Pectin-based systems improved the cytotoxicity by following ways: modulates immunological responses, induces the apoptosis, and inhibits the metastasis. The application of pectin in drug carrier is limited due to their larger particle size, low yield, and uncontrollable release kinetics. The combination of ursolic acid (UA) and 10-hydroxycampto-thecin (HCPT) loaded pectin eight-arm-PEG NPs were constricted by the self-assembly technique (Given in Figure 4.5). The stability and loading efficiency may be improved by coupling of UA and also minimized the RES uptake. The formulation improved the cellular uptake and higher survival rate due to synergistic effects than free drugs (Liu et al., 2017). In the study, pectin based nanoparticles were used to deliver the combina-tion of dihydroartemisinin (DHA) and 10-hydroxycamptothecin to the cancer cell. Using carbodiimide chemistry, pectin was conjugated with

DHA to form pectin-dihydroartemisinin (PDC) pro-drug. Then HCPT was entrapped into the PDC to form Nanoparticles. The formulation improved the loading efficacy due to the presence of a large number of carboxyl groups. The uptake studies displayed the synergistic effect as compared to free drugs. *In vivo* study demonstrated the better survival rate as compared to free drugs (Liu et al., 2017). Pectin-based colonic drug delivery is depicted in Figure 4.6.

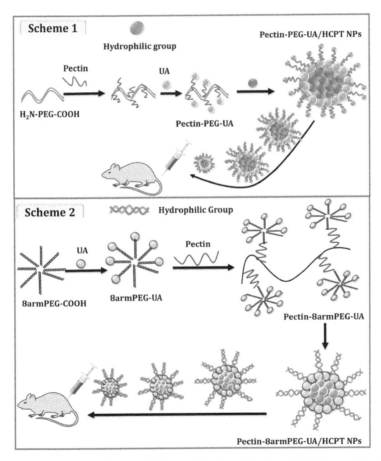

FIGURE 4.5 (See color insert.) Schematic design of the Pec-PUH NPs and Pec-8PUH NPs. Scheme 1: The Pec-PEG-UA conjugate was synthesized by introducing connecting chains (PEG), and drug molecules (UA) into pectin, and then self-assembled into Pec-PUH NPs with free HCPT being encapsulated. Scheme 2: The Pec-8armPEG-UA conjugate was synthesized by introducing connecting chains (8arm-PEG), and drug molecules (UA) into pectin, and then self-assembled into Pec-8PUH NPs with free HCPT being encapsulated.

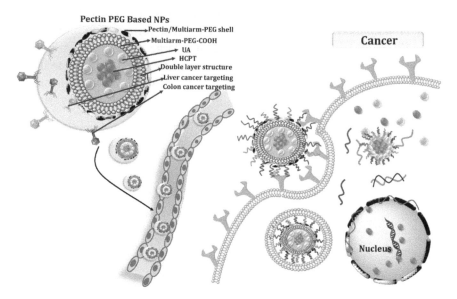

FIGURE 4.6   (See color insert.) Schematic illustration of the core-shell Pec-8PUHNPs can encapsulate both the hydrophobic drug UA and HCPT for synergistic therapy, accumulation of the nanoparticles at the tumor site through passive targeting, and specific binding to the overexpressed receptors on the tumor cells.

The pectin coating was used in order to enhance the stability of sodium caseinate/zein (CZ) complex protein NPs. pH- and heat-induced electrostatic adsorption technique was used for the coating of pectin on NPs. Pectin coating enhanced the stability of NPs significantly in GIT. The pectin coating significantly improved the capacity of NPs to encapsulate curcumin and displayed the prolong release in the gastric environment. Therefore, pectin-coated nanoparticles offer a potential platform for the oral administration of protein (Chang et al., 2017). The novel pectin based NPs were developed. The presences of galactose residues give the targeting ability for targeting the hepatocellular carcinoma. The 5-fluorouracil loaded nanoparticles were constructed using an aqueous media with the presence of calcium$^+$ and carbonate ions. The cytotoxicity study displayed the higher cytotoxic potential as compared to the free drug in HepG2 and A549 cell lines, which over-expressed asialoglycoprotein receptor (ASGPR). The pharmacokinetic study was performed in rats, and tissue distribution study was performed in mice. The results of pharmacokinetics study showed prolonged half-life and circulation as compared to the free drug (Yu et al., 2014).

4.3.4 LIPOSOMES SYSTEM

The pectin-coated nanoliposomes were prepared for targeted delivery of phloridzin. The heating-stirring-sonication technique was used for the development of liposomal formulation, and electrostatic deposition approach was used for coating the nano-liposomes with low methoxyl pectin (LMP). The pectin-coated liposomes enhanced the entrapment efficiency, storage stability, and also improved their immobilization. The cationic pectin-coated nanoliposomal formulation was the effective carrier system for the delivery of phloridzin (Haghighi et al., 2018). The dual-coated bacteriocins encapsulated liposomes were constructed for oral delivery of peptide. The negatively and positively charged liposomes were prepared from various phospholipids. Thin film casting technique was used for the preparation of formulation. Stability and release pattern of the liposomal formulation was improved by coating the liposome with two layers composed of pectin and whey proteins (WPI). The ionic interactions and weak interactions stabilized the cationic and anionic liposomal formulations, respectively. The outcomes of *in vitro* release profile demonstrated that the liposomes protect the drug into the gastric media and effectively delivered the antimicrobial peptides (Gomaa et al., 2017). The Vitamin C loaded liposomes were developed and coated with high methoxyl pectin (HMP) or LMP for transdermal application. The size was increased, and the zeta potential was decreased. FTIR study confirmed that the hydrogen bond was involved in the interaction of pectin with liposomes. The coated liposomes showed lower oxidation, leakage, and aggregation. The permeation of drug was increased up to 1.7 and 2.1 for HMP-L and LMP-L, respectively. The peptide coating enhanced the stability of liposomes, especially in the case of LMP-L (Zhou et al., 2014).

4.3.5 BEADS SYSTEM

The gellan gum and pectin are the most extensively employed polysaccharides in the pharmaceutical industry. They are low cost, biocompatible, and stable polysaccharides also have gelling properties. The gellan gum and pectin based beads were prepared for the delivery of resveratrol (RES) to the colon. The size (diameter) was 914 μm with a circular shape, entrapment efficiency was 76%, and the Span index was 0.29. The *in vitro* release profile displayed the minimum drug release in the gastric media, whereas the controlled and sustained drug release for 48 hours was observed in the intestinal pH. Weibull's model was used for calculating the release kinetics. The release kinetics followed the combination of Fickian diffusion and Case II transport. The swelling and diffusion were involved in controlled release. The formulation was safe and effective for Caco-2 and HT29-MTX intestinal cell lines. The RES loaded beads efficiently delivered the drug to the targeted colon cells and also protected the drug in the gastric medium (Prezotti et al., 2018). The carvedilol-loaded microbeads of sodium alginate (SA) were constructed by ionotropic gelation technique. The gelatin and pectin were used as release modifiers. The FTIR was used for the determination of drug-polymer interaction. The beads were characterized for various parameters like size, entrapment efficiency, swelling ratio, shape, and surface morphology by SEM, bio-adhesion study. *In vitro* release kinetics was done in 1.2N hydrochloric acid for an initial 2 hours and after that in phosphate buffer 7.4 for the remaining period of time. The outcomes showed that the microbeads had better drug loading and entrapment efficiency, low size, and moisture content. The *in vitro* release kinetics displayed the sustained drug release for 24 hours and good mucoadhesive properties (Menon and Sajeeth, 2013). The CS/ layered pectin-coated double hydroxide (LDH) beads were prepared for colonic drug delivery. The pectin-coated beads controlled the release pattern of 5-aminosalicylic acid (5-ASA) and protected the drug from the acidic pH of GIT. The *in vitro* studies displayed the controlled drug release and stability to water swelling without any influences of pH change in GIT. The composites 5-ASA loaded beads were the potential carrier for colonic drug delivery (Ribeiro et al., 2014). The residual amount of ciprofloxacin accumulated in colonic tissue may be responsible for drug resistance and killing of colonic microbiota. CS–iron microspheres encapsulated pectin beads were prepared for removing the residual colonic ciprofloxacin. The formulation absorbed the residual antibiotics and decreased the contact time with the endogenous bacteria. The Eudragit® RS coating was needed for protection in the intestinal medium. The incubation of beads was carried out in simulated gastric (SGF) and intestinal fluid (SIF) for determination of drug absorption and stability. The beads were stable in SGF and SIF for 1hr and 5 hrs, respectively (Table 4.2). The results of drug absorption revealed that the encapsulation of CS–iron in beads did not affect the absorption of CS–iron and effectively sequestrated the residual ciprofloxacin in colonic tissue (Reynaud et al., 2015).

TABLE 4.2 Various Conventional and Recent Targeted Pectin Based Carrier Systems With Their Applications

System	Drug	Composition	Application	Reference
Conventional Pectin-based system				
Tablets	Indomethacin	Calcium Chloride, Pectin	Controlled release delivery	Rubinstein et al., 1993
Tablets	Paracetamol	Ethylcellulose, pectin	Colon-specific delivery system	Wakerly et al., 1996
Matrix Tablets	Theophylline	EudragitS100 (methylacrylic acid-methylmethacrylate copolymer, soluble at pH above 7), Pectin AU 701, Pectin CF020, etc.	pH-sensitive colon delivery	Mura et al., 2003
Matrix Tablets	Theophylline	Pectin type CU020, CU201, CU501, CU701, etc.	Controlled release colon delivery	Sriamornsak et al., 2007
Novel targeted pectin based carrier systems				
Beads	Resveratrol	Pectin, gellan gum	Colonic drug delivery	Prezotti et al., 2018
	Carvedilol	Sodium alginate, pectin, and calcium Chloride, Gelatin	Prolong release and better mucoadhesion	Menon and Sajeeth, 2013
Nanoparticles	Itraconazole (ITZ)	low methoxyl pectin (LMP), amidated low methoxyl pectin (ALMP), or high methoxyl pectin (HMP)	Antifungal drug delivery	Burapapadh et al., 2012
	Curcumin	Sodium caseinate/zein (CZ), pectin	Oral delivery vehicles.	Chang et al., 2017
	Ursolic acid and 10-hydroxycamptothecin	Eight-arm-polyethylene glycol carboxylic acid, pectin	Anticancer drug delivery	Liu et al., 2017
Hydrogel	Mannitol	Pectin, gelatin, and glutaraldehyde	Stimuli-responsive/ pH-responsive drug delivery system	Naeem et al., 2017

TABLE 4.2 *(Continued)*

System	Drug	Composition	Application	Reference
	5-fluorouracil (5-FU)	Pectin, poly ((2-dimethylamino) ethyl methacrylate)) and phosphate crosslinkerbis[2-methacryloyloxy] ethyl phosphate	Dual responsive targeting both as anticancer as well as antibacterial targeting	Eswaramma et al., 2017
	Hyaluronic acid		In the field of cartilage tissue engineering.	Chen et al., 2017
	Simvastatin	Sodium alginate and Pectin/Alginate	For the use in the field of biomedical technology and in wound dressings	Rezvanian et al., 2017
	soybean hull peroxidase	Soybean peroxidase, glycidyl-methacrylate, periodate, glutaraldehyde, and pectin	For the purpose of increased thermal and organic solvent stability	Prokopijevic et al., 2017
	Ketoprofen lysinate (KL)	Alginate/Pectin	Use for the oral administration of NSAIDs in chronic inflammatory diseases	Cerciello et al., 2016
Microsphere	Vancomycin	Chitosan, mucin, pectin, and pectinase enzyme.	Colon-specific delivery system	Bigucci et al., 2009
	Ketoprofen and benzoic acid	Starch, alginate, and pectin via supercritical CO_2-assisted adsorption	For oral drug delivery	García-González et al., 2015
	Ibuprofen	Tricalcium phosphate, sodium glycerophosphate, and pectin	Use for generating macroporous; Bone cements and for developing controlled release properties	Fullana et al., 2010
	Prednisolone	Pectin, Prednisolone, and Eudragit	For the treatment of ulcerative colitis/colon targeting system	Dashora and Jain, 2009

TABLE 4.2 *(Continued)*

System	Drug	Composition	Application	Reference
	Ondansetron	Pectin LM (Low Molecular Weight) and Ondansetron hydrochloride.	For nasal drug delivery system	Mahajan et al., 2012
	Aceclofenac	Eudragit S-100 and pectin	For the treatment of rheumatoid arthritis	Ramasamy et al., 2013
	Salicylic acid	Pectin, $CaCl_2$, and Plasticizer	For the preparation of Controlled drug delivery system	Kistriyani et al., 2016
	Metformin hydrochloride	Pectin, Ethylcellulose, Liquid paraffin light, Hydrochloric acid, Sodium hydroxide, Span-80, Acryl coat S 100, etc.	For good drug entrapment and release kinetics	Sanjoy et al., 2013
	Ambroxol hydrochloride	Low methoxy pectin, Croscarmellose, Microcrystalline cellulose, Polyethylene glycol 1000, etc.	For masking the bitterness of the drug as well as for the preparation of fast disintegrating dosage forms	Jacob and Shirwaikar, 2009
	Ciprofloxacin hydrochloride	Pectin and chitosan	For the treatment of osteomyelitis.	Orhan et al., 2006
	Acetylsalicylic acid	Alginate- pectin matrix, chitosan, and $CaCl_2$	For the targeted drug release at intestine	Jaya et al., 2010

4.4   CONCLUSION

Among the natural polymers, pectin is extensively utilized in the field of drug delivery prospects because of their number of useful and unique properties for the development of various pectin based carrier systems for a vista of applications using new technology. Pectin-based systems have shown great promise for controlled delivery of varied therapeutic agents, including colon specific targeting. These pectin-based drug delivery systems can be produced on a large scale with ease and cost-effectiveness. Recent advances in terms of improved safety and efficacy along with the implementation of other suitable polymers and modifications may endow multiple characteristics and rediscover new insights. This chapter encompasses the basic knowledge about pectin and its application potential in various fields of drug delivery as well as the conventional and recent pectin based targeted carrier systems with principal fabrication techniques and characterization techniques. Moreover, there is a need to take care of regulatory aspects and clinical concern for achieving the fullest potential of all investigated new recent pectin-based drug carrier systems.

KEYWORDS

- **composites biocompatibility**
- **drug delivery**
- **hetero-polysaccharide**
- **natural polymer**
- **pectin**

REFERENCES

Ahrabi, S. F., Madsen, G., Dyrstad, K., Sande, S. A., & Graffner, C., (2000). Development of pectin matrix tablets for colonic delivery of model drug ropivacaine. *Eur. J. Pharm. Sci.*, *10*(1), 43–52.

Almeida, E. A., Facchi, S. P., Martins, A. F., Nocchi, S., Schuquel, I. T., Nakamura, C. V., Rubira, A. F., & Muniz, E. C., (2015). Synthesis and characterization of pectin derivative with antitumor property against Caco-2 colon cancer cells. *Carbohydr. Polym.*, *115*, 139–145.

Ashford, M., Fell, J. T., Attwood, D., Sharma, H., & Woodhead, P. J., (1993). An *in vivo* investigation into the suitability of pH-dependent polymers for colonic targeting. *Int. J. Pharm.*, *95*(1–3), 193–199.

Bauer, H. G., Asp, N. G., Dahlqvist, A., Fredlund, P. E., Nyman, M., & Öste, R., (1981). Effect of two kinds of pectin and guar gum on 1, 2-dimethylhydrazine initiation of colon tumors and on fecal β-glucuronidase activity in the rat. *Cancer Res.*, *41*(6), 2518–2523.

Bigucci, F., Luppi, B., Monaco, L., Cerchiara, T., & Zecchi, V., (2009). Pectin-based microspheres for colon-specific delivery of vancomycin. *J. Pharm. Pharmacol.*, *61*(1), 41–46.

Burapapadh, K., Takeuchi, H., & Sriamornsak, P., (2012). Novel pectin-based nanoparticles prepared from nanoemulsion templates for improving *in vitro* dissolution and *in vivo* absorption of poorly water-soluble drug. *Eur. J. Pharm. Biopharm.*, *82*(2), 250–261.

Byrd, J. C., & Bresalier, R. S., (2004). Mucins and mucin binding proteins in colorectal cancer. *Cancer Metastasis Rev.*, *23*(1/2), 77–99.

Cerciello, A., Auriemma, G., Del, G. P., Cantarini, M., & Aquino, R. P., (2016). Natural polysaccharides platforms for oral controlled release of ketoprofen lysine salt. *Drug Dev. Ind. Pharm*, *42*(12), 2063–2069.

Chang, C., Wang, T., Hu, Q., Zhou, M., Xue, J., & Luo, Y., (2017). Pectin coating improves physicochemical properties of caseinate/zein nanoparticles as oral delivery vehicles for curcumin. *Food Hydrocolloids*, *70*, 143–151.

Chatterjee, B., Amalina, N., Sengupta, P., & Mandal, U. K., (2017). Mucoadhesive polymers and their mode of action: A recent update. *Journal of Applied Pharmaceutical Science, 7*(05), 195–203.

Chen, F., Ni, Y., Liu, B., Zhou, T., Yu, C., Su, Y., Zhu, X., Yu, X., & Zhou, Y., (2017). Self-crosslinking and injectable hyaluronic acid/RGD-functionalized pectin hydrogel for cartilage tissue engineering. *Carbohydr. Polym.*, *166*, 31–44.

Chen, J., Park, H., & Park, K., (1999). Synthesis of superporous hydrogels: Hydrogels with fast swelling and superabsorbent properties. *J. Biomed. Mater. Res. A.*, *44*(1), 53–62.

Chen, W., Li, H., Shi, D., Liu, Z., & Yuan, W., (2016). Microneedles as a delivery system for gene therapy. *Front Pharmacol.*, *7*, 137.

Cheng, H., Li, S., Fan, Y., Gao, X., Hao, M., Wang, J., Zhang, X., Tai, G., & Zhou, Y., (2011). Comparative studies of the antiproliferative effects of ginseng polysaccharides on HT-29 human colon cancer cells. *Med. Oncol.*, *28*(1), 175–181.

Chung, W. S. F., Meijerink, M., Zeuner, B., Holck, J., Louis, P., Meyer, A. S., Wells, J. M., Flint, H. J., & Duncan, S. H., (2017). Prebiotic potential of pectin and pectic oligosaccharides to promote anti-inflammatory commensal bacteria in the human colon. *FEMS Microbiol. Ecol.*, *93*(11), 127.

Da Costa, M. P. M., De Mello, F. I. L., & De Macedo, C. M. T., (2016). New polyelectrolyte complex from pectin/chitosan and montmorillonite clay. *Carbohydr. Polym.*, *146*, 123–130.

Dafe, A., Etemadi, H., Dilmaghani, A., & Mahdavinia, G. R., (2017). Investigation of pectin/starch hydrogel as a carrier for oral delivery of probiotic bacteria. *Int. J. Biol. Macromol.*, *97*, 536–543.

Dashora, A., & Jain, C., (2009). Development and characterization of pectin-prednisolone microspheres for colon targeted delivery. *Int. J. Chem. Tech. Res.*, *1*(3), 751–757.

Demir, Y. K., & Kerimoglu, O., (2015). Novel use of pectin as a microneedle base. *Chem. Pharm. Bull., (Tokyo)*, *63*(4), 300–304.

Demir, Y. K., Metin, A. Ü., Şatıroğlu, B., Solmaz, M. E., Kayser, V., & Mäder, K., (2017). Poly (methyl vinyl ether-co-maleic acid)–Pectin based hydrogel-forming systems: Gel, film, and microneedles. *Eur. J. Pharm. Biopharm.*, *117*, 182–194.

Do Prado, S. B. R., Ferreira, G. F., Harazono, Y., Shiga, T. M., Raz, A., Carpita, N. C., & Fabi, J. P., (2017). Ripening-induced chemical modifications of papaya pectin inhibit cancer cell proliferation. *Sci. Rep.*, *7*(1), 16564.

Drury, J. L., & Mooney, D. J., (2003). Hydrogels for tissue engineering: Scaffold design variables and applications. *Biomaterials*, *24*(24), 4337–4351.

Eltayib, E., Brady, A. J., Caffarel-Salvador, E., Gonzalez-Vazquez, P., Alkilani, A. Z., McCarthy, H. O., McElnay, J. C., & Donnelly, R. F., (2016). Hydrogel-forming microneedle arrays: Potential for use in minimally-invasive lithium monitoring. *Eur. J. Pharm. Biopharm.*, *102*, 123–131.

Eswaramma, S., Reddy, N. S., & Rao, K. K., (2017). Phosphate crosslinked pectin based dual responsive hydrogel networks and nanocomposites: Development, swelling dynamics and drug release characteristics. *Int. J. Biol. Macromol.*, *103*, 1162–1172.

Ferguson, L. R., & Harris, P. J., (1996). Studies on the role of specific dietary fibers in protection against colorectal cancer. *Mutat. Res. Fund. Mol. Mech. Mut.*, *350*(1), 173–184.

Fullana, S. G., Ternet, H., Freche, M., Lacout, J. L., & Rodriguez, F., (2010). Controlled release properties and final macroporosity of a pectin microspheres–calcium phosphate composite bone cement. *Acta Biomater.*, *6*(6), 2294–2300.

García-González, C. A., Jin, M., Gerth, J., Alvarez-Lorenzo, C., & Smirnova, I., (2015). Polysaccharide-based aerogel microspheres for oral drug delivery. *Carbohydr. Polym.*, *117*, 797–806.

Glinsky, V. V., & Raz, A., (2009). Modified citrus pectin anti-metastatic properties: One bullet, multiple targets. *Carbohydr. Res.*, *344*(14), 1788–1791.

Gomaa, A. I., Martinent, C., Hammami, R., Fliss, I., & Subirade, M., (2017). Dual coating of liposomes as encapsulating matrix of antimicrobial peptides: Development and characterization. *Frontiers in Chemistry*, 5.

Haghighi, M., Yarmand, M. S., Emam-Djomeh, Z., McClements, D. J., Saboury, A. A., & Rafiee-Tehrani, M., (2018). Design and fabrication of pectin-coated nanoliposomal delivery systems for a bioactive polyphenolic: Phloridzin. *Int. J. Biol. Macromol.*, *112*, 626–637.

Heitman, D., Hardman, W., & Cameron, I., (1992). Dietary supplementation with pectin and guar gum on 1, 2-dimethylhydrazine-induced colon carcinogenesis in rats. *Carcinogenesis*, *13*(5), 815–818.

Hodges, L., Connolly, S., Band, J., O'Mahony, B., Ugurlu, T., Turkoglu, M., Wilson, C., & Stevens, H., (2009). Scintigraphic evaluation of colon targeting pectin–HPMC tablets in healthy volunteers. *Int. J. Pharm.*, *370*(1/2), 144–150.

Hong, X., Wu, Z., Chen, L., Wu, F., Wei, L., & Yuan, W., (2014). Hydrogel microneedle arrays for transdermal drug delivery. *Nano-Micro Letters*, *6*(3), 191–199.

Jacob, S., & Shirwaikar, A., (2009). Preparation and evaluation of microencapsulated fast melt tablets of ambroxol hydrochloride. *Indian J. Pharm. Sci.*, *71*(3), 276.

Jaya, S., Durance, T., & Wang, R., (2010). Physical characterization of drug loaded microcapsules and controlled in vitro release study. *The Open Biomaterials Journal*, *2*(1), 9–17.

Jiang, Y., Du, Y., Zhu, X., Xiong, H., Woo, M. W., & Hu, J., (2012). Physicochemical and comparative properties of pectins extracted from Akebia trifoliata var. australis peel. *Carbohydr. Polym.*, *87*(2), 1663–1669.

Kim, B., & Peppas, N. A., (2003). Poly (ethylene glycol)-containing hydrogels for oral protein delivery applications. *Biomed. Microdevices.*, *5*(4), 333–341.

Kistriyani, L., Wirawan, S., & Sediawan, W., (2016). In: *Effect of ca2+ to Salicylic Acid Release in Pectin Based Controlled Drug Delivery System* (p. 012042). IOP conference series: Materials science and engineering, IOP Publishing.

Liu, L., Fishman, M. L., Kost, J., & Hicks, K. B., (2003). Pectin-based systems for colon-specific drug delivery via oral route. *Biomaterials*, *24*(19), 3333–3343.

Liu, Y., Liu, K., Li, X., Xiao, S., Zheng, D., Zhu, P., Li, C., Liu, J., He, J., & Lei, J., (2017). A novel self-assembled nanoparticle platform based on pectin-eight-arm polyethylene glycol-drug conjugates for co-delivery of anticancer drugs. *Mater. Sci. Eng. C.*, *86*, 28–41.

Lupton, J. R., (2000). Is fiber protective against colon cancer? Where the research is leading us. *Nutrition*, *16*(7), 558–561.

Mahajan, H. S., Tatiya, B. V., & Nerkar, P. P., (2012). Ondansetron loaded pectin based microspheres for nasal administration: *In vitro* and *in vivo* studies. *Powder Technol.*, *221*, 168–176.

Marras-Marquez, T., Peña, J., & Veiga-Ochoa, M., (2015). Robust and versatile pectin-based drug delivery systems. *Int. J. Pharm.*, *479*(2), 265–276.

May, C. D., (1990). Industrial pectins: Sources, production and applications. *Carbohydr. Polym.*, *12*(1), 79–99.

Menon, T. V., & Sajeeth, C., (2013). Formulation and evaluation of sustained release sodium alginate microbeads of Carvedilol. *Int. J. Pharmtech. Res.*, *5*(2), 746–753.

Mishra, R., Banthia, A., & Majeed, A., (2012). Pectin-based formulations for biomedical applications: A review. *Asian J. Pharm. Clin. Res.*, *5*(4), 1–7.

Moore, M. A., Park, C. B., & Tsuda, H., (1998). Soluble and insoluble fiber influences on cancer development. *Crit. Rev. Oncol. Hematol.*, *27*(3), 229–242.

Morita, R., Honda, R., & Takahashi, Y., (2000). Development of oral controlled release preparations, a PVA swelling-controlled release system (SCRS): I. Design of SCRS and its release controlling factor. *J. Control Release*, *63*(3), 297–304.

Munarin, F., Guerreiro, S., Grellier, M., Tanzi, M., Barbosa, M., Petrini, P., & Granja, P., (2011). Pectin-based injectable biomaterials for bone tissue engineering. *Biomacromolecules*, *12*(3), 568–577.

Mura, P., Maestrelli, F., Cirri, M., Luisa, G. R. M., & Rabasco, A. A. M., (2003). Development of enteric-coated pectin-based matrix tablets for colonic delivery of theophylline. *J. Drug Target*, *11*(6), 365–371.

Naeem, F., Khan, S., Jalil, A., Ranjha, N. M., Riaz, A., Haider, M. S., Sarwar, S., Saher, F., & Afzal, S., (2017). pH-responsive cross-linked polymeric matrices based on natural polymers: Effect of process variables on swelling characterization and drug delivery properties. *BioImpacts: BI*, *7*(3), 177.

Niu, R., Qin, Z., Ji, F., Xu, M., Tian, X., Li, J., & Yao, F., (2017). Hybrid pectin–Fe 3+/ polyacrylamide double network hydrogels with excellent strength, high stiffness, superior toughness and notch-insensitivity. *Soft Matter*, *13*(48), 9237–9245.

Noreen, A., Akram, J., Rasul, I., Mansha, A., Yaqoob, N., Iqbal, R., Tabasum, S., Zuber, M., & Zia, K. M., (2017). Pectins functionalized biomaterials, a new viable approach for biomedical applications: A review. *Int. J. Biol. Macromol.*, *101*, 254–272.

Olano-Martin, E., Rimbach, G. H., Gibson, G. R., & Rastall, R. A., (2002). Pectin and pectic-oligosaccharides induce apoptosis in *in-vitro* human colonic adenocarcinoma cells. *Anticancer Res.*, *23*(1A), 341–346.

Orhan, Z., Cevher, E., Mülazimoglu, L., Gürcan, D., Alper, M., Araman, A., & Özsoy, Y., (2006). The preparation of ciprofloxacin hydrochloride-loaded chitosan and pectin microspheres: Their evaluation in an animal osteomyelitis model. *Bone Joint J.*, *88*(2), 270–275.

Osada, Y., Okuzaki, H., & Hori, H., (1992). A polymer gel with electrically driven motility. *Nature*, *355*(6357), 242.

Paharia, A., Yadav, A. K., Rai, G., Jain, S. K., Pancholi, S. S., & Agrawal, G. P., (2007). Eudragit-coated pectin microspheres of 5-fluorouracil for colon targeting. *AAPS Pharm. Sci. Tech.*, *8*(1), 12.

Park, Y., Park, J., Chu, G. S., Kim, K. S., Sung, J. H., & Kim, B., (2015). Transdermal delivery of cosmetic ingredients using dissolving polymer microneedle arrays. *Biotechnol. Bioprocess Eng.*, *20*(3), 543–549.

Prezotti, F. G., Boni, F. I., Ferreira, N. N., Campana-Filho, S. P., Almeida, A., Vasconcelos, T., Gremião, M. P. D., Cury, B. S. F., & Sarmento, B., (2018). Gellan gum/pectin beads are safe and efficient for the targeted colonic delivery of resveratrol. *Polymers*, *10*(1), 50.

Prokopijevic, M., Prodanovic, O., Spasojevic, D., Kovacevic, G., Polovic, N., Radotic, K., & Prodanovic, R., (2017). Tyramine-modified pectins via periodate oxidation for soybean hull peroxidase induced hydrogel formation and immobilization. *Appl. Microbiol. Biotechnol.*, *101*(6), 2281–2290.

Ramasamy, T., Ruttala, H. B., Shanmugam, S., & Umadevi, S. K., (2013). Eudragit-coated aceclofenac-loaded pectin microspheres in chronopharmacological treatment of rheumatoid arthritis. *Drug Deliv.*, *20*(2), 65–77.

Reynaud, F., Tsapis, N., Guterres, S. S., Pohlmann, A. R., & Fattal, E., (2015). Pectin beads loaded with chitosan–iron microspheres for specific colonic adsorption of ciprofloxacin. *J. Drug Deliv. Sci. Technol.*, *30*, 494–500.

Rezvanian, M., Ahmad, N., Amin, M. C. I. M., & Ng, S. F., (2017). Optimization, characterization, and in vitro assessment of alginate-pectin ionic cross-linked hydrogel film for wound dressing applications. *Int. J. Biol. Macromol.*, *97*, 131–140.

Ribeiro, L. N., Alcântara, A. C., Darder, M., Aranda, P., Araújo-Moreira, F. M., & Ruiz-Hitzky, E., (2014). Pectin-coated chitosan–LDH bionanocomposite beads as potential systems for colon-targeted drug delivery. *Int. J. Pharm.*, *463*(1), 1–9.

Rubinstein, A., & Radai, R., (1995). *In vitro* and *in vivo* analysis of colon specificity of calcium pectinate formulations. *Eur. J. Pharm. Biopharm.*, *41*(5), 291–295.

Rubinstein, A., Radai, R., Ezra, M., Pathak, S., & Rokem, J. S., (1993). *In vitro* evaluation of calcium pectinate: A potential colon-specific drug delivery carrier. *Pharm. Res.*, *10*(2), 258–263.

Sanjoy, K., Swagata, D. R., & Bhusan, S. H., (2013). Development & optimization of pectin microsphere of metformin HCL. *Int. J. Drug Dev.*, *5*(1), 339–356.

Sriamornsak, P., (2003). Chemistry of pectin and its pharmaceutical uses: A review. *SUIJ*, *3*(1/2), 206–228.

Sriamornsak, P., Sungthongjeen, S., & Puttipipatkhachorn, S., (2007). Use of pectin as a carrier for intragastric floating drug delivery: Carbonate salt contained beads. *Carbohydr. Polym.*, *67*(3), 436–445.

Thibault, J. F., & Ralet, M. C., (2003). Physico-chemical properties of pectins in the cell walls and after extraction. In: *Advances in Pectin and Pectinase Research* (pp. 91–105). Springer.

Tiwary, A., & Rana, V., (2016). Utilizing pectin interactions for modifying drug release. *Future Science*, *7*(8), 517–520.

Turkoglu, M., & Ugurlu, T., (2002). *In vitro* evaluation of pectin–HPMC compression coated 5-aminosalicylic acid tablets for colonic delivery. *Eur. J. Pharm. Biopharm.*, *53*(1), 65–73.

Ugurlu, T., Turkoglu, M., Gurer, U. S., & Akarsu, B. G., (2007). Colonic delivery of compression coated nisin tablets using pectin/HPMC polymer mixture. *Eur. J. Pharm. Biopharm.*, *67*(1), 202–210.

Umar, S., Morris, A., Kourouma, F., & Sellin, J., (2003). Dietary pectin and calcium inhibit colonic proliferation *in vivo* by differing mechanisms. *Cell Prolif.*, *36*(6), 361–375.

Vaidya, A., Jain, S., Agrawal, R. K., & Jain, S. K., (2015). Pectin–metronidazole prodrug bearing microspheres for colon targeting. *J. Saudi Chem. Soc.*, *19*(3), 257–264.

Wakerly, Z., Fell, J., Attwood, D., & Parkins, D., (1996). Pectin/ethylcellulose film coating formulations for colonic drug delivery. *Pharm. Res.*, *13*(8), 1210–1212.

Wang, W., Wu, X., Chantapakul, T., Wang, D., Zhang, S., Ma, X., Ding, T., Ye, X., & Liu, D., (2017). Acoustic cavitation-assisted extraction of pectin from waste grapefruit peels: A green two-stage approach and its general mechanism. *Food Res. Int.*, *102*, 101–110.

Wang, X., Chen, Q., & Lü, X., (2014). Pectin extracted from apple pomace and citrus peel by subcritical water. *Food Hydrocolloids*, *38*, 129–137.

Yadav, P., Pandey, P., & Parashar, S., (2017). Pectin as natural polymer: An overview. *RJPT*, *10*(4), 1225–1229.

Yu, C. Y., Wang, Y. M., Li, N. M., Liu, G. S., Yang, S., Tang, G. T., He, D. X., Tan, X. W., & Wei, H., (2014). *In vitro* and *in vivo* evaluation of pectin-based nanoparticles for hepatocellular carcinoma drug chemotherapy. *Mol. Pharm.*, *11*(2), 638–644.

Zhang, W., Mahuta, K. M., Mikulski, B. A., Harvestine, J. N., Crouse, J. Z., Lee, J. C., Kaltchev, M. G., & Tritt, C. S., (2016). Novel pectin-based carriers for colonic drug delivery. *Pharm. Dev. Technol.*, *21*(1), 127–130.

Zhang, W., Xu, P., & Zhang, H., (2015). Pectin in cancer therapy: A review. *Trends Food Sci. Technol.*, *44*(2), 258–271.

Zhou, W., Liu, W., Zou, L., Liu, W., Liu, C., Liang, R., & Chen, J., (2014). Storage stability and skin permeation of vitamin C liposomes improved by pectin coating. *Colloids Surf B Biointerfaces*, *117*, 330–337.

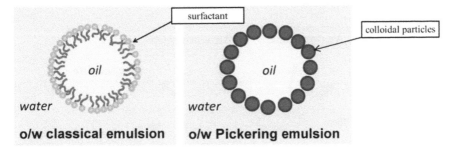

FIGURE 2.1 Structure diagram of Pickering emulsion and conventional emulsion. Modified with permission from Chevalier and Bolzinger, 2013. © Elsevier.

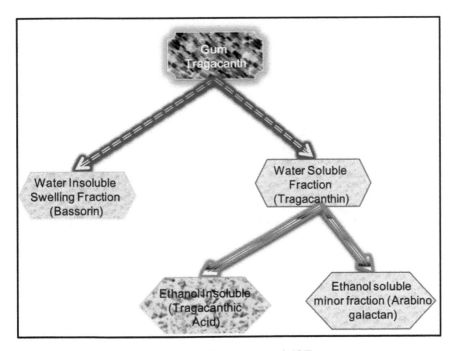

FIGURE 3.1 Fractional composition of gum tragacanth (GT).

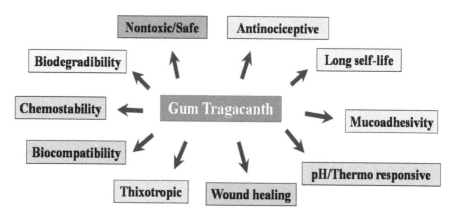

FIGURE 3.3   Characteristic properties of gum tragacanth (GT).

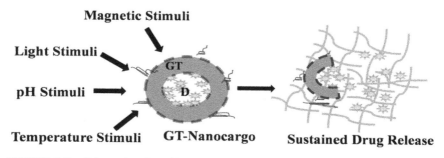

FIGURE 3.5   Schematic of multi-stimuli responsive GT-encapsulated/hydrogel based site-specific drug delivery system (GT – Gum Tragacanth, D – desired drug).

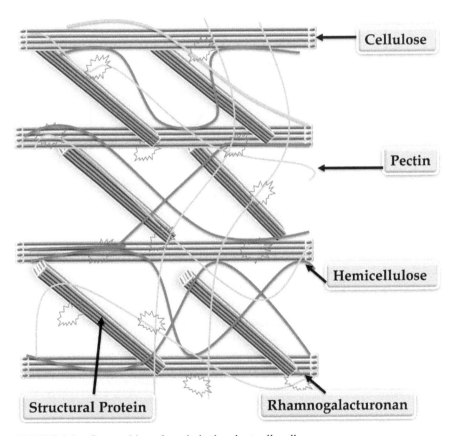

FIGURE 4.1   Composition of pectin in the plant cell wall.

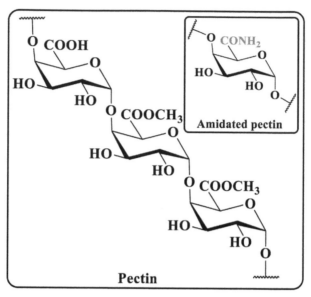

FIGURE 4.2   Chemical structures of pectin and amidated pectin.

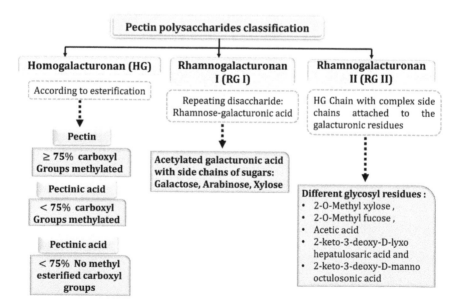

FIGURE 4.3   Classification of pectin.

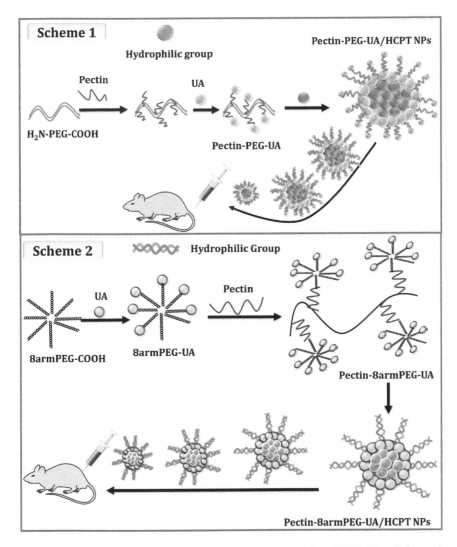

FIGURE 4.5 Schematic design of the Pec-PUH NPs and Pec-8PUH NPs. Scheme 1: The Pec-PEG-UA conjugate was synthesized by introducing connecting chains (PEG), and drug molecules (UA) into pectin, and then self-assembled into Pec-PUH NPs with free HCPT being encapsulated. Scheme 2: The Pec-8armPEG-UA conjugate was synthesized by introducing connecting chains (8arm-PEG), and drug molecules (UA) into pectin, and then self-assembled into Pec-8PUH NPs with free HCPT being encapsulated.

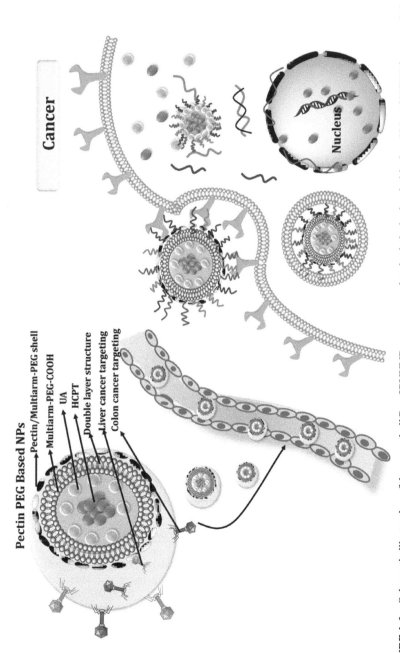

FIGURE 4.6 Schematic illustration of the core-shell Pec-8PUHNPs can encapsulate both the hydrophobic drug UA and HCPT for synergistic therapy, accumulation of the nanoparticles at the tumor site through passive targeting, and specific binding to the overexpressed receptors on the tumor cells.

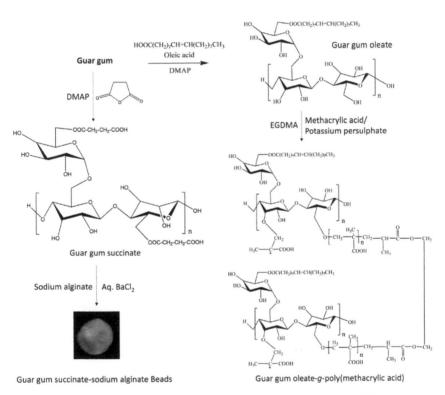

FIGURE 5.5   Scheme for the preparation of guar gum-based colon drug delivery carriers.

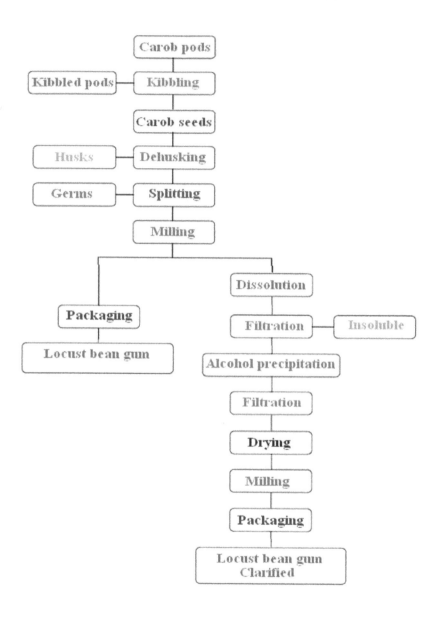

General flow chart of LBG extraction

FIGURE 6.1   General flow chart of LBG extraction.

CHAPTER 5

Guar Gum and Its Derivatives: Pharmaceutical Applications

D. SATHYA SEELI and M. PRABAHARAN

Department of Chemistry, Hindustan Institute of Technology and Science, Padur, Chennai – 603103, India

ABSTRACT

Guar gum is a non-ionic natural polymer. Because of the presence of enormous hydroxyl groups, guar gum can be chemically converted into a variety of derivatives with the improved physicochemical properties. Over the past few decades, guar gum has been widely considered as potential biomaterials due to its hydrophilicity, non-toxicity, biocompatibility, and biodegradability. This chapter is aimed at focusing the biomedical applications of guar gum-based materials.

5.1 INTRODUCTION

Guar gum is obtained from the seeds of *Cyamopsistetragonolobus* (Jain et al., 2017; Giri et al., 2017). It contains $(1{\rightarrow}4)$-β-D-mannopyranosyl and α-D-galactopyranosyl units connected by $(1{\rightarrow}6)$ linkages as represented in Figure 5.1 (Moreira and Filho, 2008). The galactose side chains are arbitrarily linked to a mannose unit with a ratio of 1:2 in the building blocks of guar gum (Tripathy and Das, 2013).

Guar gum can be modified into a variety of guar gum derivatives with different functionalities as potential biomaterials for biomedical uses (Jani et al., 2009; Dziadkowiec et al., 2017; Iqbal, 2013). In recent years, guar gum has been extensively studied as a prospective material for colon drug delivery due to its sustained drug release property, stability and vulnerability to bacterial degradation in the gastrointestinal tract (GIT) (Bayliss and Houston, 1986; Vinaykumar et al., 2011; Tomolin et al., 1989). Moreover, guar gum derivatives have also been examined as scaffolds for tissue engineering due to their biodegradability, biocompatibility, ability to form a porous structure, absorption of biological fluids, adequate mechanical stability and positive interactions with cells (Tiwari et al., 2009). This chapter intends to overview the recent progress and applications of guar gum-based materials in drug delivery, tissue engineering, wound healing, and biosensor fields.

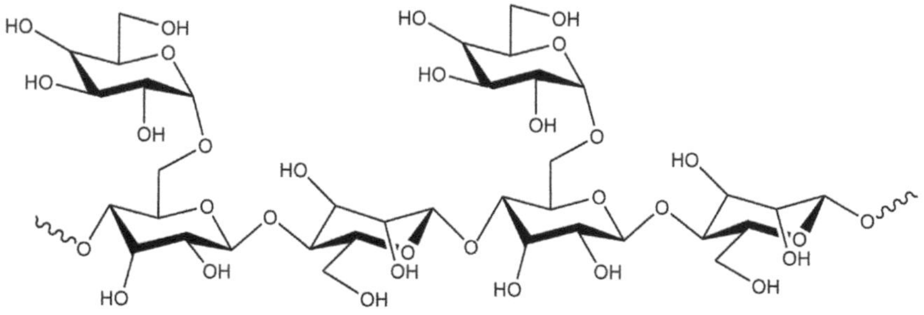

FIGURE 5.1   Structure of g\uar gum.

5.2   PHARMACEUTICAL APPLICATIONS

5.2.1   *COLON DRUG DELIVERY*

Guar gum-based materials have been extensively studied as colon-specific drug delivery carriers because of their desired characteristics and vulnerability to microbial actions in the colonic environment.

5.2.1.1   *GUAR GUM-BASED TABLETS*

Guar gum-based tablets containing therapeutic agents have shown promising results in the treatment of colon diseases. Wong et al., (1997) developed the guar gum-based tablets loaded with dexamethasone and budesonide. These tablets showed a minimum amount of drug release in replicated gastric and intestinal media. However, they presented a maximum release of drug in simulated colonic fluid (SCF). This study confirmed that the galactoman-nanase present in SCF accelerated the release of the encapsulated drugs from the guar gum tablets. The extent of drug release was found to be influenced by the concentration of galactomannanase. Indomethacin (IND)-loaded guar gum tablets were developed by Ramaprasad et al., (1998). These tablets were found to stable in the both acidic and mild alkaline medium for about 2–3 h. The amount of IND released from the tablets was found to be about 21% in 2% (w/v) cecal medium, while it was about 91% in 4% (w/v) cecal medium.

Krishnaiah et al., (1999) assessed the guar gum as a covering material to defend 5- aminosalicylic acid in the upper GIT. In simulated gastric fluid (SGF), the tablets covered with guar gum (150–300mg) presented a sustained drug release for 26 h. The drug coated with 300 and 200 mg of guar gum presented 23.85 and 63.43% of drug release in the medium, respectively for 26 h. These observations suggested that coating of drug with guar gum plays an important role in delivering the drugs to the colonic site. Krishnaiah et al., (2001) also prepared 5- aminosalicylic acid and mebendazole tablets coated with guar gum. The drug release studies showed that the tablets comprising about 20–30% of guar gum could be suitable to deliver the mebendazole in the colonic site.

The guar gum tablets containing the mixture of albendazole, metroni-dazole, and tinidazole were prepared for the simultaneous administration of drugs to the colon (Krishnaiah et al., 2001a). It was observed that the release of albendazole from the tablets was reduced when increasing the administrated amount of metronidazole and tinidazole. In another study, different types of metronidazole tablets coated with different contents of guar gum were developed for delivery of colon drugs (Krishnaiah et al., 2002). Based on the amount of guar gum present in the system, the matrix and multilayer tablets released ~52 and 44% of the drug, respectively in the GIT. The compression coated tablets delivered <1% of the drug in the stomach and small intestine region. Whereas in SCF, these tablets released another 61% of the drug in 24 h due to the disintegration by colonic bacteria. These results revealed the suitability of these tablets for the delivery of drugs in the colon site.

Intravenous 5-fluorouracil administration could provide severe side-effects because of its toxicity to the normal cells. To overcome this issue, tablets based on guar gum were prepared to release the drugs more precisely to the colon site. Krishnaiah et al., (2002b) developed 5-fluorouracil tablets covered with various amounts of guar gum and studied their drug release behaviors by a high-performance liquid chromatography (HPLC) technique. In the SGF, these tablets released ~2.5–4% of the drug. However, these tablets presented higher amount of drug release in the SCF. The tablets covered with 80% of guar gum were found to be ideal for colon cancer therapy as they exhibited less amount of drug release in SGF. The guar gum tablets consist of mebendazole were reported for the treatment of colon disease (Krishnaiah et al., 2003). During the oral administration, the drug was delivered in a sustained way from the tablets and started increasing its concentration to 25.7 ng/ml in 9.4 h, indicated that guar gum coated mebendazole tablets could limit the drug release in the stomach and enhance the substantial amount of drug release in the colonic site.

Sinha et al., (2004) developed guar gum tablets with a reduced coating weight for the delivery of 5-fluorouracil. Accordingly, 5-fluorouracil tablets covered with a mixture of xanthan/guar gum and studied their drug release characters. The amount of drug release from the tablets comprising of xanthan and guar gum in the ratio of 20:20, 20:10 and 10:20 was found to be ~18%, 20%, and 30%, respectively after 24 h of suspension. Studies on compression coated tablets containing 10:20 ratio of xanthan and guar gum showed 67.2% and 80.34% of 5-fluorouracil release in 2% and 4% (w/v) of the cecal content medium, respectively after 19 h of incubation. A matrix tablet composed of guar gum, xanthan gum, and tamarind gum (TmG) loaded with betulinic acid was developed by Murali Kishan (2014). This matrix tablet displayed an improved drug release behavior when compared to control tablet made by guar gum alone. Using a wet granulation method, guar gum-based matrix tablets containing budesonide were prepared by Vivekanandan et al., (2015). The results exhibited that the release of budesonide from the tablets was influenced by the swelling and degradation behaviors of guar gum in the colonic environment. The developed tablets showed ~97.12% of drug release in rat cecalmedium. However, these tablets were found to release only ~76.86% of the drug in the dissolution medium without cecal content.

Hashem et al., (2016) evaluated the suitability of guar gum/hydroxypropyl methylcellulose (HPMC) coating for the release of prednisolone. The drug release profile revealed that the rate of prednisolone release was improved with an increase in HPMC concentration up to 20% at pH 7.4 medium consist of 2% (w/v) rat cecal content. The *in vivo* study showed that the developed tablets completely disintegrated in the colonic region while maintaining their integrity in the stomach and small intestine. Recently, a mixture of guar gum and pectin coated matrix tablets with 5% Eudragit® L-100 of quinazolinone derivative were developed for colon site-specific delivery (Shammika et al., 2017). It was found that all the prepared formulations of matrix tablets showed a less amount of drug release in an acidic medium for 2 h and faster release in the mild alkaline medium for another 3 h. The results showed that the tablets based on guar gum/pectin/Eudragit® L-100 of quinazolinone derivative could be more effective in releasing the drug into the colon.

5.2.1.2   CROSSLINKED GUAR GUM

In recent years, guar gum-based materials crosslinked with different crosslinking agents have gained much importance as colon drug delivery carriers. Gliko-Kabir et al., (2000, 2000a) prepared guar gum hydrogels crosslinked with phosphate groups. The drug release behavior of hydrocortisone loaded guar gum hydrogel showed that the phosphate crosslinking within the hydrogel resisted 80% of the drug release at pH 6.4 medium for 6 h. It was found that the amount of drug release was improved when incorporating the enzymes such as α-galactosidase and β-mannanase in the release medium. *In vivo* release studies displayed the controlled degradation of guar gum by the enzymes depending upon their concentration. Using the identical approach, Chourasia et al., (2004, 2006) described the glutaraldehyde (GA) crosslinked guar gum microspheres for colon drug delivery. These microspheres were found to maintain a similar drug release pattern in both PBS and SGF. However, these particles exhibited the maximum amount (91.0%) of drug release in the buffer consists of rat cecal content. *In vivo* release study showed that the quantity of drug delivered to the colon from the microspheres was higher than that of control drug suspension.

Kaushik et al., (2016) developed GA crosslinked guar gum microspheres by emulsion polymerization method as 5-aminosalicylic acid delivery carriers for the management of colon disease. The amount of 5-aminosalicylic acid released from the microspheres was found to be only 13.93% in SGF, and simulated intestinal fluid (SIF) for 5 h, which indicated the higher stability of GA crosslinked guar gum microsphere in the SGF and SIF medium. Carboxymethyl guar gum nanoparticles with an average diameter of 208 nm were also developed by ionic gelation method for colon drug delivery (Dodi et al., 2016). The nanoparticles loaded with rhodamine-B showed a pH-dependent drug release behavior in simulated gastrointestinal (GI) fluids. The MTT assay revealed that the carboxymethyl guar gum nanoparticles crosslinked with trisodium trimetaphosphate are non-toxic up to its concentrations ~0.3 mg/ml.

5.2.1.3   GUAR GUM DERIVATIVES

The increased water uptake behavior of guar gum may limit its application as drug delivery carriers. To overcome this issue, chemical modification of guar gum would be a suitable approach. Using the chemical change, the swelling ability of guar gum can be restricted. Moreover, the desired properties can be introduced to guar gum for the efficient release of drugs to the colon site through oral administration. Soppimath et al., (2001) developed the spherically shaped guar gum-g-poly (acrylamide) microgels by the emulsification technique. These microgels showed a response to pH and ionic strength of medium because of the existence of carboxylic acid groups. The drugs released from the microgels were found to be obeyed the pH-dependent swelling followed by the relaxation-controlled mechanism.

Soppimath et al., (2002) also prepared GA crosslinked guar gum-g-poly (acrylamide) hydrogel for the release of verapamil hydrochloride and nifedipine, as shown in Figure 5.2. The drug release from the hydrogel was found to be largely influenced by drug loading content, crosslinking density, nature of the drug and drug loading technique. In another study, Toti and Aminabhavi (2004) synthesized guar gum-g-poly (acrylamide) by changing the composition of guar gum and acrylamide. Here, the amide groups of grafted polymer were changed to carboxylic groups to introduce the pH-responsive characters to the resulting drug carriers.

Sen et al., (2010) prepared guar gum-g-poly (acrylamide) hydrogel for the delivery of 5-aminosalicylic acid as depicted in Figure 5.3. The results showed that the rate of drug release from hydrogel was decreased with an increase in grafting percentage. Moreover, the rate of drug delivery was determined to be lower at acidic medium than that at neutral and alkaline media. This result suggested that guar gum-g-poly (acrylamide) can be suitable for colon-targeted delivery of drugs. Shahid et al., (2013) also prepared guar gum-g-poly (acrylamide) by microwave irradiation method. In this study, the microstructure of prepared material was examined with XRD and SEM analysis. It was observed that the crystallinity of the grafted material was improved and the granular structure of guar gum was altered

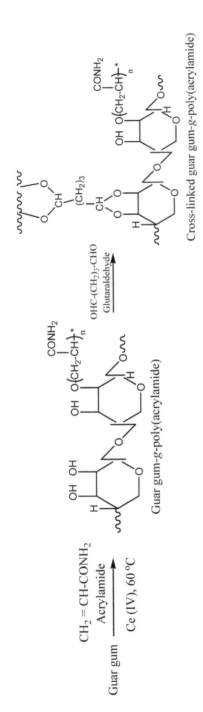

FIGURE 5.2 Formation of crosslinked guar gum-*g*-poly (acrylamide).

into a fibrillar arrangement due to poly (acrylamide) grafting. The grafted material showed a relationship between the microwave exposure time and drug release pattern.

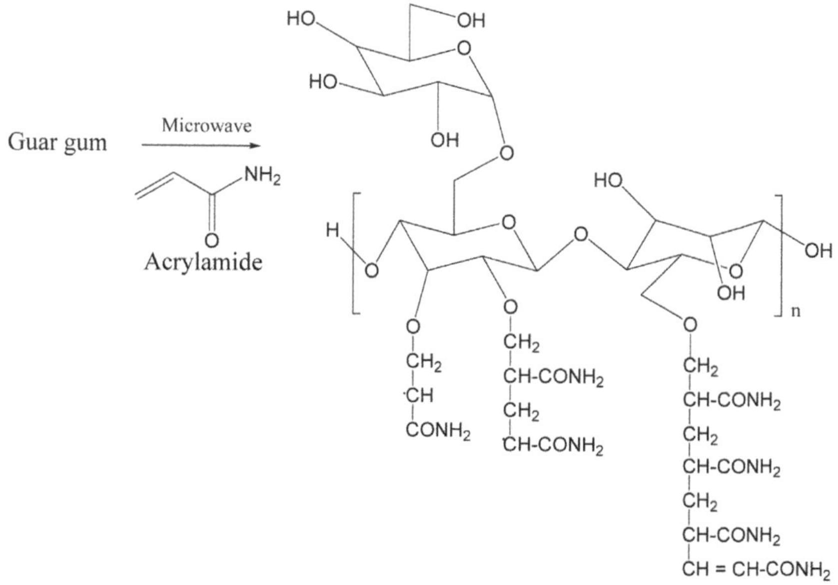

FIGURE 5.3   Preparation of guar gum-*g*-poly (acrylamide).

Panda et al., (2014) formulated calcium chloride crosslinked microbeads based on guar gum-*g*-poly (acrylamide) blended with sodium alginate (SA). These microbeads were found to have the spherical shape as indicated by SEM studies. The swelling index and doxofyllinerelease of the microbeads were higher at pH 7.4 than at pH 1.2 media. The microbeads with increased poly (acrylamide) and SA content presented the sustained release of the encapsulated drug. The *in vitro* release kinetics of doxofylline from the polymeric beads followed Higuchi kinetics model. Varma et al., (2016) reported pH-sensitive esomeprazole magnesium-loaded guar gum-*g*-poly (acrylamide) nanocarriers for colon drug delivery. The sizes of the nanocarriers were found to be about 200–600 nm. The encapsulation and drug loading abilities of these materials were determined as 33.2–50.1% and 12.2–17.2%, respectively. It was found that the particle size and drug-loading efficacy of the nanoparticles were increased with an increase in copolymer concentration. The drug release experiments conducted at pH

1.2 and 6.8 revealed a pH-sensitive drug release behavior of the developed nanocarrier. This nanocarrier limited the initial drug release at acidic pH and followed the controlled drug release over a prolonged period at alkaline pH.

Li et al., (2006) developed guar gum-*g*-poly (acrylic acid (AAc)) based hydrogels and studied their swelling kinetics and water transport mechanism. The swelling behavior of these hydrogels was found to be influenced by their chemical composition and environmental pH. Huang et al., (2007) developed the polyelectrolyte hydrogels using the cationic guar gum-*g*-poly (AAc) for colon drug delivery. In this study, cationic guar gum was prepared by reacting 3-chloro-2-hydroxypropyl trimethylammonium chloride with guar gum in the presence of NaOH (Figure 5.4). The swelling and drug release behavior of the hydrogels were found to be influenced by the pH of the environments, preparative methods, and composition of hydrogels. At pH 7.4, the higher amount of drug release was observed from the hydrogel, which indicated that polymer chain relaxation plays a vital role in the drug release mechanism.

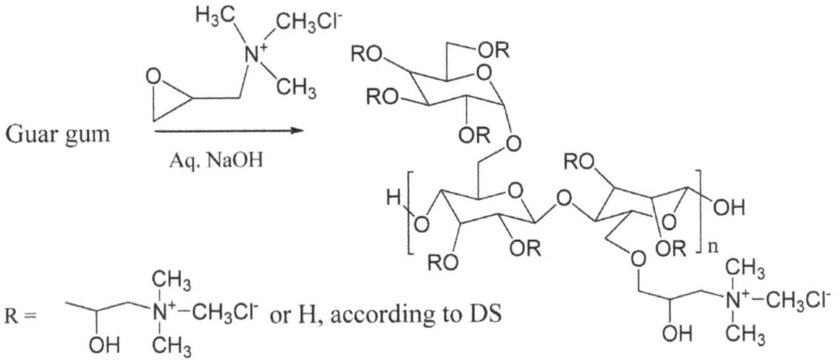

FIGURE 5.4 Preparation of cationic guar gum.

Thakur et al., (2009) prepared different types of AAc grafted guar gum hydrogels using free radical polymerization. These hydrogels presented the pH and ionic strength-dependent swelling behavior. To analyze the drug release profile, these hydrogels were encapsulated with L-tyrosine and L-3, 4-dihydroxyphenylalanine (L-DOPA). The release studies conducted at acidic and mild basic medium showed that the maximum amount of L-tyrosine and L-DOPAwas released from the hydrogels based on guar gum-*g*-poly (methacrylic acid) and guar gum-*g*-poly (AAc), respectively.

The hydrogel-based on guar gum-g-poly (2-hydroxypropyl methacrylate) showed the sustained drug release and retained about 50% of the drug within the polymer matrix even after 13 h.

Sharma et al., (2013) developed GA crosslinked microspheres composed of guar gum and methacrylic acid. These microspheres presented the pH-dependent water uptake and drug delivery characters in the buffer solutions. These microspheres were found to follow a swelling-controlled drug release mechanism in the SIF. The hydrogels composed of guar gum, AAc, and β-cyclodextrin (β-CD)were developed using a crosslinking agent tetraethyl orthosilicate for intestinal delivery of dexamethasone (Dasand Subuddhi, 2015). Due to the presence of β-CD, the developed hydrogels showed the sustained release of the encapsulated drug when compared to the control. In addition, it was noted that the amount of drug release was influenced by the guar gum content present in the hydrogel network. When the increasing concentration of guar gum, the rate of drug release was decreased, and the release period was extended. These results indicated that the hydrogels containing β-CD could be ideal for the oral delivery of dexamethasone to the colon. MTT assay showed that these hydrogels are non-toxic and biocompatible. Recently, Sathya Seeliand Prabaharan (2016a, b; 2017) developed the guar gum-based pH-sensitive colon drug delivery carriers, as shown in Figure 5.5. These materials presented varying swelling and drug release behaviors based on the pH of the medium. These materials presented a maximum amount of drug release at mild alkaline medium (pH 7.4) than that at acidic medium (pH 1.2). Because of the presence of a crosslinked network within the polymer matrix, the developed materials had improved stability at both acidic and alkaline buffer medium for more than 30 h. Cytotoxicity studies revealed that the prepared guar gum-based materials had no obvious cytotoxicity against the mouse mesenchymal stem cells.

5.2.2 *ANTIHYPERTENSIVE DRUG DELIVERY*

Guar gum-based materials have been considered for the delivery of antihypertensive drugs (Friend et al., 1997; Altaf et al., 1998). The guar gum matrices showed the controlled release of ketoprofen, nifedipine, and diltiazem hydrochloride in the simulated body fluid. Krishnaiah et al., (2002a) prepared the matrices based on guar gum for the delivery of trimetazidine dihydrochloride through the oral route. The release studies indicated that these matrix tablets could be used for oral delivery of hydrophilic drugs in a sustained manner. Totiand Aminabhavi (2004) developed the matrices based on guar gum-*g*-poly (acrylamide) and hydrolyzed guar gum-*g*-poly (acrylamide) for diltiazem hydrochloride delivery. These materials presented the controlled release of the loaded drug for more than 8 h. The hydrolyzed guar gum-*g*-poly (acrylamide) matrix presented about 73% of the drug release in the SIF.

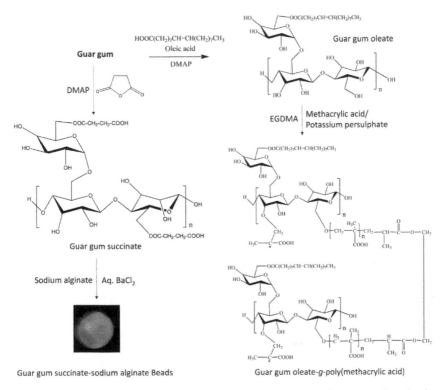

FIGURE 5.5 **(See color insert.)** Scheme for the preparation of guar gum-based colon drug delivery carriers.

Al Saidan et al., (2005) prepared the guar gum matrix for the oral administration of diltiazem hydrochloride. These tablets showed the controlled drug release through the Fickian-diffusion mechanism in SGF and SCF when compared to commercial diltiazem hydrochloride tablets. These observations confirmed that guar gum-based matrix tablets could be suitable for antihypertensive drug delivery. Soumya et al., (2010) prepared the nanoparticles based on lipase functionalized guar gum as antihypertensive drug carriers. These nanoparticles showed a higher amount of drug release up to 24h and then presented the less amount of release in the later period. This result proposed that the nanocarriers based on guar gum could be ideal for antihypertensive drug delivery. Recently, Panigrahy et al., (2016) reported the formulation of atenolol floating-bioadhesive tablets for the treatment of hypertension. These tablets were developed with compression technique using the polymer composites consisting of guar gum, HPMC, carbopol, and SA. In this study, the prepared tablets were evaluated for the buoyancy test, mucoadhesion force, swelling study, drug content, *in vitro* release profile and *ex vivo* mucoadhesion strength.

5.2.3 TRANSDERMAL DRUG DELIVERY

Transdermal drug delivery systems could be more effective to deliver the drugs to the bloodstream in a controlled manner and maintain their concentration for a longer period (Fathima et al., 2017). Over the past few years, different types of guar gum-based materials have been developed for transdermal drug delivery because of their film-forming capability, improved hydrophilicity, and controlled drug release property. Murthy et al., (2004) developed the transdermal drug delivery systems based on terbutaline sulfate-loaded carboxymethyl guar gum. Since carboxymethyl guar gum exhibits enough film-forming capability, it has been used to make the films with the desired swelling and controlled drug release properties by changing their concentrations. Giri et al., (2013) prepared the guar gum-*g*- AAc/nanosilica membranes for the diclofenac sodium (DS) delivery (Figure 5.6). These membranes displayed the controlled release of DS because of their hydrophobic interactions with the drug molecules. Giri et al., (2014) also developed the DS-loaded composite membranes based on multi-walled carbon nanotube (MWCNT) functionalized carboxymethyl guar gum-*g*-2-hydroxyethyl methacrylate. These membranes presented the less sustained and more swelling controlled drug release. Nevertheless, they showed the improved half-life period of encapsulated drugs.

Chattopadhyay et al., (2016) formulated glucosamine sulfate-loaded guar gum/sodium carboxymethyl cellulose systems for the treatment of osteoarthritis. The optimum concentrations of guar gum, carboxymethyl cellulose, and glycerol required for the development of 25 g of transdermal gel were found to be 419 mg, 445 mg, and 2322 mg, respectively. The optimized transdermal gel showed the controlled release of glucosamine sulfate by following the Fickian release mechanism. The *in vivo* study conducted by using rat model indicated the deviations in the release profile of the proinflammatory cytokine tumor necrosis factor-α, which demonstrated the effectiveness of glucosamine sulfate-loaded guar gum/sodium carboxymethyl cellulose systems for the therapy of osteoarthritis. Recently, Anirudhan et al., (2017) constructed the transdermal device that made up from cationic guar gum and borate modified poly (vinyl alcohol) (PVA) polyelectrolyte complex reinforced with nanogold-nanocellulose composites for the administration of diltiazem hydrochloride. This device showed improved thermomechanical property, film transparency, antimicrobial activity, and cell viability.

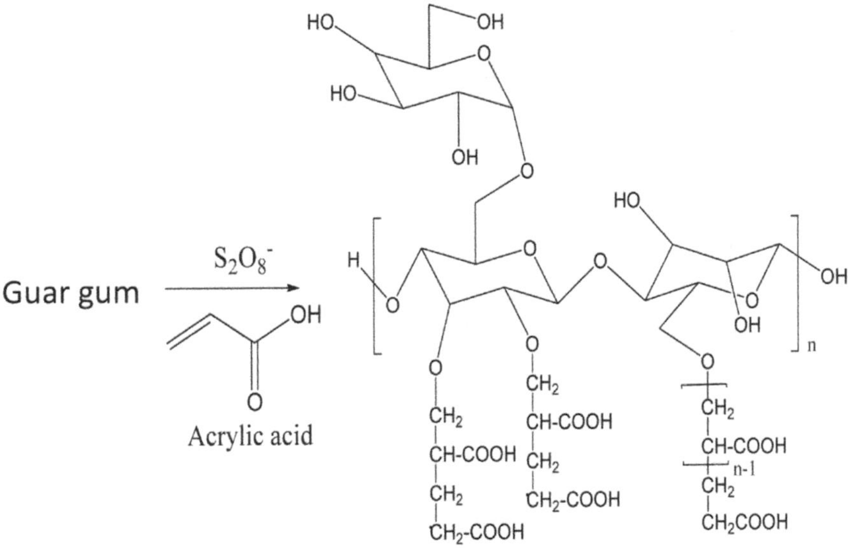

FIGURE 5.6　Reaction scheme for the preparation of acrylic acid (AAc) grafted guar gum.

5.2.4　TISSUE ENGINEERING SCAFFOLDS

The main goal of tissue engineering is to construct the perfect scaffolds that can offer suitable milieu for the reconstruction of tissues (Prabaharan and Jayakumar, 2009; Jafari et al., 2017). Finding the appropriate materials, along with suitable cells, could play an important role in the growth of ideal scaffolds for tissue engineering. In recent years, natural polymers like alginate, chitosan (CS), guar gum and hyaluronic acid have received a great interest in the designing of varies scaffolds (Hong et al., 2007; Iwasaki et al., 2004; Rafat et al., 2008; Smith et al., 2006; Kuo and Ma, 2001). Being a non-ionic biopolymer, guar gum has a large scope to be utilized as scaffolds for tissue engineering. Zhao et al., (2006) developed the hydroxypropyl guar gum/poly (ethylene glycol)-*co*-poly (ε-caprolactone) diacrylate hydrogels as scaffolds using UV irradiation technique and studied their swelling, mechanical, and BSA releases properties. Due to the semi-IPN, these hydrogels presented better swelling, mechanical, and controlled drug release behaviors. Tiwari et al., (2009) prepared the guar gum-*g*-methacrylate hydrogels by photopolymerization method (Figure 5.7). These hydrogels showed a three-dimensional microstructure and an equilibrium swelling ratio of 22 to 63%. The enzymatic degradation of the hydrogels was found to be decreased with an increase in hydrogel concentration and extent of methacrylation. MTT assay showed that the developed hydrogels are non-toxic to the human endothelial cell line.

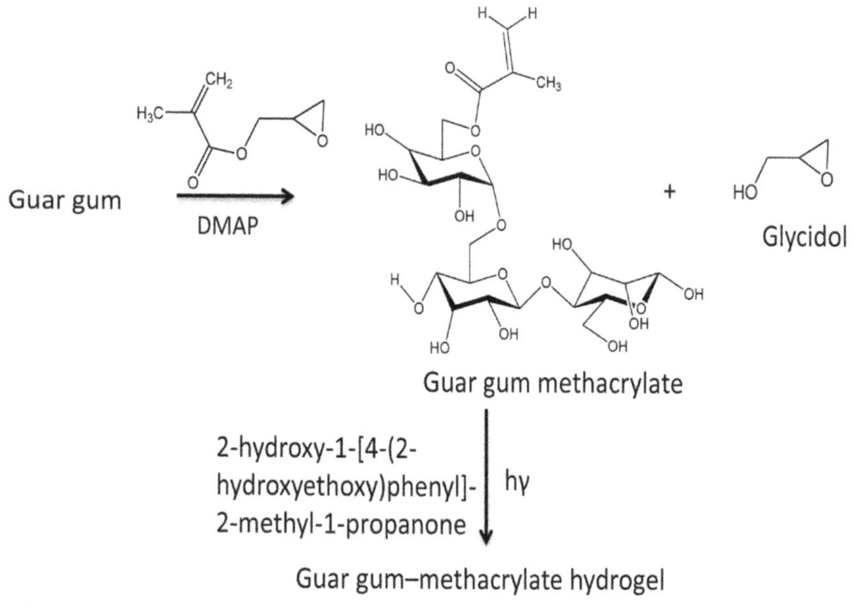

FIGURE 5.7   Preparation of guar gum-*g*-methacrylate hydrogel.

The impact of guar gum on the physical properties of collagen-based scaffolds was assessed by Manikoth et al., (2012). Guar gum was found to alter the rheological behavior of the collagen without changing its basic protein structure. Guar gum/collagen composites presented the smaller pores, which enabled effective cell-polymer interaction. When increasing the content of guar gum, the thermal stability of the scaffolds was increased. Moreover, the polarizability of the collagen was altered by varying the amount of guar gum in the composites. Ragothaman et al., (2014) developed the tissue engineering scaffolds based on guar gum and collagen/poly (dialdehyde) hybrid materials. The biodegradability, swelling ability, and thermal behaviors of these hybrid scaffolds were found to be improved because of the formation of cross-linked collagen networks within the hybrid scaffolds. These scaffolds showed the sustained delivery of immobilized platelet and increased fibroblast cell density and proliferation. Murali et al., (2016) confirmed that the collagen/poly (dialdehyde) guar gum hybrid scaffolds encapsulated with platelet could stimulate the chemotactic effects in the rooted spot to enhance quick wound repair and tissue regeneration without the aid of antibacterial agents.

5.2.5 WOUND HEALING MATERIALS

Wound healing is an inborn biological rejoinder, which aids reinstate cellular and anatomic steadiness of a tissue. A perfect wound healing material should be non-toxic, non-allergenic, non-adherent, easily detached and made by appropriate biomaterials that retain adequate wetness at the wound boundary and allow a gaseous exchange (Jayakumar et al., 2011). As a compatible biopolymer, the growth of guar gum as wound healing materials has witnessed in recent years. Auddy et al., (2013) prepared a wound healing material based on cationic guar gum alkylamine loaded with silver nanoparticles. It was observed that the silver nanoparticles were dispersed well within the polymer matrix. Hence, these nanoparticles enhanced the antibacterial activity and wound healing process when compared to commercial material. The contents of DNA, total protein, and hydroxyproline were increased in the cells treated with cationic guar gum alkylamine loaded with silver nanoparticles. In addition, it was noted that the cationic guar gum derivative offered a moistures surface required for cell adhesion and proliferation.

Sevinc et al., (2014) prepared the partially hydrolyzed guar gum derivatives and studied their colonic anastomotic healing effects using the rat models. The results of this study indicated that colonic anastomotic healing was unfavorably influenced by pre-operative radio-treatment. However, orogastric feeding with hydrolyzed guar gum improved the healing procedure. Pramanik et al., (2015) designed the composite films based on guar gum/polyhydroxyalkanoates/curcumin blend as a wound healing and antibacterial material. Due to the improved homogeneity and surface roughness because of the presence of curcumin, the developed films showed the enhanced attachment and proliferation of cells. Jana et al., (2016) developed a ceftazidime-loaded film based on the polymer blend consists of aminated carboxymethyl guar gum and fish scale collagen. These films showed an appreciable antimicrobial effect on *Staphylococcus aureus* and *Pseudomonas aeruginosa*. The thermal stability and the mechanical property of these films were found to be enhanced when compared to the carboxymethyl guar gum and native collagen. The results of biocompatibility and blood compatibility studies suggested that this film can be efficiently used for wound healing application. Recently, a nanoemulsion of menthol/peppermint oil/methyl salicylate-loaded guar gum hydrogels was found to be suitable for symptomatic relief in patients with intractable pruritus and colonic wound treatment (Wu et al., 2016). Horii et al., (2016) found that partially hydrolyzed guar gum fibers can be effective in the healing of intestinal mucosal epithelium by enhancing the activation of RhoA (A protein-coding gene).

5.2.6   BIOSENSORS

The major components of the biosensor are biological responsive component and physical transducer. It converts the biological recognition incident into an appropriate physical signal. In recent years, various types of biopolymers combined with enzyme immobilize electrodes have been reported for the identification of biomolecules (Prabaharan, 2013; Deng et al., 2016). Since the enzyme activity plays a significant role in the performance of biosensors, a great interest is devoted for finding the suitable matrix that can hold the enzymes and retain their biological activities (Kushwah and Bhadauria, 2010). In this respect, the materials based on guar gum have a great scope to be considered as an immobilization matrix for biomolecules due to their desired physicochemical properties.

Bagal and Karve (2006) developed the invertase enzyme-immobilized porous membranes based on agarose/guar gum composite and studied their sucrose hydrolytic activity. The immobilization efficacy of invertase enzyme was determined as 91%. Moreover, the activity of invertase was found to be good at pH 4.5–6.5. The enzyme presented the twelve cycles of reusability, improved thermal, and operational stability due to its conformational stability on the surface of the agarose/guar gum matrix. Using a similar approach, Tembe et al., (2006, 2007) developed an agarose/guar gum matrix immobilized with tyrosinase for the identification of micromolar level L-DOPA and dopamine.

Bagal et al., (2007) prepared agarose/guar gum matrix loaded with acid invertase and glucose oxidase for recognition of sucrose. This sensor showed improved operational and storage stability and response to the sucrose solutions at different concentrations under the broad acidic pH ranges. Kestwal et al., (2008) prepared the agarose-guar gum matrix immobilized with invertase and glucose oxidase for the detection of sucrose. This matrix showed the detection limit about 1×10^{-10} to 1×10^{-7} M at pH 5.5. Pandey et al., (2012) prepared a nanocomposite based on guar gum/silver nanoparticles and analyzed its ammonia sensing property by an optical technique. The developed nanocomposite presented the ammonia detection limit up to 1 ppm and response time about 2–3 seconds. Vaghela et al., (2014) developed the processable polyaniline film by *in situ* polymerizations of agarose-guar gum assisted aniline for the construction of chemical and biosensors. This composite film showed considerable electrochemical activity and direct current conductivity due to its electro-responsive character. Dhananjayan et al., (2017) reported a biosensor and pseudo capacitor based on polypyrrole/ cerium-nickel modified guar gum nanocomposite. In this study, polypyrrole, and cerium-nickel bimetallic nanospheres after irradiated with the electron beam were functionalized into amine functionalized guar gum film. This electrode exhibited an excellent electrocatalytic activity towards the sensing of hydroquinone, resorcinol, catechol, and nitrite.

5.3 CONCLUDING REMARKS

Guar gum-based materials have received a great interest in drug delivery, tissue engineering, wound healing, and biosensors due to their favorable properties, namely non-toxicity, biodegradability, and biocompatibility. The guar gum derivatives presented better drug loading ability, sustained drug release property, stimuli-responsive behavior, stability, and site specificity due to the existence of required functionalities. Crosslinked guar gum-based materials were found to have increased stability and prolonged period of sustained drug release. The pH-sensitive guar gum derivatives have great utility in the field of colon-specific protein delivery because of their pH-responsive swelling and drug release behaviors. Due to the better film-forming capability, guar gum-based systems have a wide scope to be utilized as transdermal drug delivery systems. Guar gum-based matrices can be used as tissue engineering scaffolds due to their increased water uptake and network mesh size. Guar gum-based materials with non-toxic, biocompatible, non-adherent, and antimicrobial properties were found to promote the wound healing process. Since guar gum matrix immobilized with enzyme retains its specific biological function, it is considered for biosensor applications.

ACKNOWLEDGMENT

The authors are grateful to DST-Nano Mission, Department of Science and Technology, India, for financial assistance through grant SR/NM/NS-1260/2013.

KEYWORDS

- **anti-microbial activities**
- **biosensors**
- **drug delivery**
- **guar gum**
- **tissue engineering**

REFERENCES

Al Saidan, S. M., et al., (2005). *In vitro* and *in vivo* evaluation of guar gum matrix tablets for oral controlled release of water-soluble diltiazem hydrochloride. *AAPS Pharm. Sci. Tech., 6*, 14–21.

Altaf, S. A., et al., (1998). Guar gum-based sustained release diltiazem. *Pharm. Res., 15,* 1196–1201.

Anirudhan, T. S., Nair, S. S., & Sekhar, V. C., (2017). Deposition of gold-cellulose hybrid nanofiller on a polyelectrolyte membrane constructed using guar gum and poly(vinyl alcohol) for transdermal drug delivery. *J. Memb. Sci., 539,* 344–357.

Auddy, R. G., et al., (2013). New guar biopolymer silver nanocomposites for wound healing applications. *Bio. Med. Res. Int.,* 912458.

Bagal, D. S., et al., (2007). Fabrication of sucrose biosensor based on single mode planar optical waveguide using co-immobilized plant invertase and GOD. *Biosens. Bioelectron, 22,* 3072–3079.

Bagal, D., & Karve, M. S., (2006). Entrapment of plant invertase within the novel composite of agarose – guar gum biopolymer membrane. *Anal. Chim. Acta, 555,* 316–321.

Bayliss, C. E., & Houston, A. P., (1986). Characterization of plant polysaccharide-mucin-fermenting anaerobic bacteria from human feces. *Appl. Environ. Microbiol., 48,* 626–632.

Chattopadhyay, H., et al., (2016). Accentuated transdermal application of glucosamine sulfate attenuates experimental osteoarthritis induced by monosodium iodoacetate. *J. Mater. Chem. B, 4*(25), 4470–4481.

Chaurasia, M., et al., (2006). Crosslinked guar gum microspheres: A viable approach for improved delivery of anticancer drugs for the treatment of colorectal cancer. *AAPS Pharm. SciTech., 7*(3), E143–E151.

Chourasia, M. K., and Jain, S. K., Potential of guar gum microspheres for target specific drug release to colon. *J. Drug Target,* 2004, *12,* 435–442.

Das, S., & Subuddhi, U., (2015). pH-responsive guar gum hydrogels for controlled delivery of dexamethasone to the intestine. *Int. J. Bio. Macromol., 79,* 856–863.

Deng, J., Liang, W., & Fang, J., (2016). Liquid crystal droplet-embedded biopolymer hydrogel sheets for biosensor applications. *ACS Appl. Mater. Interfaces, 8*(6), 3928–3932.

Dhananjayan, N., et al., (2017). Stable and robust nanobiocomposite preparation using aminated guar gum (mimic activity of graphene) with electron beam irradiated polypyrrole and Ce-Ni bimetal: Effective role in simultaneous sensing of environmental pollutants and pseudo capacitor applications. *Electrochim Acta, 246,* 484–496.

Dodi, G., et al., (2016). Carboxymethyl guar gum nanoparticles for drug delivery applications: Preparation and preliminary *in-vitro* investigations. *Mater. Sci. Eng. C., 63,* 628–636.

Dziadkowiec, J., et al., (2017). Preparation, characterization and application in controlled release of ibuprofen-loaded guar gum/montmorillonite bionanocomposites. *Appl. Clay Sci., 135,* 52–63.

Fathima, S. A., Begum, S., & Fatima, S. S., (2017). Transdermal drug delivery system. *Int. J. Pharm. Clin. Res., 9*(1), 35–43.

Friend, D. R., et al., (1997). Development of a zero-order nifedipine dosage form using COSRx technology. *Proc. Int. Sym. Control. Rel. Bioact. Mat., 24,* 311–312.

Giri, A., Bhunia, T., Mishra, S. R., Goswami, L., Panda, A. B., Pald, S., & Bandyopadhyay, A., (2013). Acrylic acid grafted guar gum–nanosilica membranes for transdermal diclofenac delivery. *Carbohydr. Polym., 91*(2), 492–501.

Giri, A., et al., (2014). A transdermal device from 2-hydroxyethyl methacrylate grafted carboxymethyl guar gum–multi-walled carbon nanotube composites. *RSC Advances, 4*(26), 13546–13556.

Giri, S., et al., (2017). Effect of addition of enzymatically modified guar gum on glycemic index of selected Indian traditional foods. *Bioact. Carbohydr. Diet. Fiber, 11,* 1–8.

Gliko-Kabir, I., et al., (2000a). Phosphated crosslinked guar for colon-specific drug delivery: I. Preparation and physicochemical characterization. *J. Control. Release, 63,* 121–127.

Gliko-Kabir, I., et al., (2000b). Phosphated crosslinked guar for colon-specific drug delivery: II. *In vitro* and *in vivo* evaluation in the rat. *J. Control. Release, 63,* 129–134.

Hong, Y., et al., (2007). Covalently crosslinked chitosan hydrogel: Properties of *in vitro* degradation and chondrocyte encapsulation. *J. Shen. Acta Biomater., 3,* 23–31.

Horii, Y., et al., (2016). Partially hydrolyzed guar gum enhances colonic epithelial wound healing: Via activation of RhoA and ERK1/2. *RSC Food Funct., 7*(7), 3176–3183.

Huang, Y., Yu, H., & Xiao, C., (2007). pH-sensitive cationic guar gum/poly (acrylic acid) polyelectrolyte hydrogels: Swelling and *in vitro* drug release. *Carbohydr. Polym., 69*(4), 774–783.

Iqbal, D. N., (2013). Synthesis and characterization of guar gum derivatives with antioxidant moieties. *Int. J. Pharm. Bio. Sci., 4*(4), 305–316.

Iwasaki, N., et al., (2004). Feasibility of polysaccharide hybrid materials for scaffolds in cartilage tissue engineering: Evaluation of chondrocyte adhesion to polyion complex fibers prepared from alginate and chitosan. *Biomacromolecules, 5,* 828–33.

Jafari, M., et al., (2017). Polymeric scaffolds in tissue engineering: A literature review. *J. Biomed. Mater. Res. B Appl. Biomater., 105*(2), 431–459.

Jain, V., et al., (2017). Guar gum as a selective flocculant for the beneficiation of alumina-rich iron ore slimes: Density functional theory and experimental studies. *Minerals Eng., 109,* 144–152.

Jana, P., et al., (2016). Preparation of guar gum scaffold film grafted with ethylenediamine and fish scale collagen, crosslinked with ceftazidime for wound healing application. *Carbohydr. Polym., 153,* 573–581.

Jani, K. G., et al., (2009). Gums and mucilages: Versatile excipients for pharmaceutical formulations. *Asian J. Pharm. Sci., 4,* 308–322.

Jayakumar, R., et al., (2011). Biomaterials based on chitin and chitosan in wound dressing applications. *Biotechnol. Adv., 29,* 322–337.

Kaushik, D., Sharma, K., & Sardana, S., (2016). Colon targeting guar gum microspheres of 5-aminosalicylic acid: Evaluation of various process variables, characterization and *in-vitro* drug release. *Ind. J. Pharm. Edu. Research, 50*(2), 106–114.

Kestwal, D. B., et al., (2008). Invertase inhibition based electrochemical sensor for the detection of heavy metal ions in aqueous system: Application of ultra-microelectrode to enhance sucrose biosensor's sensitivity. *Biosens. Bioelectron, 24,* 657–664.

Krishnaiah, Y. S. R., et al., (2001a). Guar gum as a carrier for colon-specific delivery, influence of metronidazole and tinidazole on in vitro release of albendazole from guar gum matrix tablets. *J. Pharm. Pharmaceut. Sci., 4*(3), 235–243.

Krishnaiah, Y. S. R., et al., (2001b). Development of colon targeted drug delivery systems for mebendazole. *J. Control. Release, 77,* 87–95.

Krishnaiah, Y. S. R., et al., (2002a). Three-layer guar gum matrix tablet formulations for oral controlled delivery of highly soluble trimetazidine dihydrochloride. *J. Control. Release, 81,* 45–56.

Krishnaiah, Y. S. R., et al., (2002b). In vitro drug release studies on guar gum-based colon targeted oral drug delivery systems of 5-fluorouracil. *Eur. J. Pharm. Sci., 16,* 185–192.

Krishnaiah, Y. S. R., et al., (2002c). Studies on the development of oral colon targeted drug delivery systems for metronidazole in the treatment of amoebiasis. *Int. J. Pharm., 236,* 43–55.

Krishnaiah, Y. S. R., et al., (2003). Pharmacokinetic evaluation of guar gum-based colon targeted drug delivery systems of mebendazole in healthy volunteers. *J. Control. Release, 88,* 95–103.

Krishnaiah, Y. S. R., Satyanaryana, S., & Rama, P. Y. V., (1999). Studies of guar gum compression-coated 5-aminosalicylic acid tablets for colon-specific drug delivery. *Drug. Dev. Ind. Pharm., 25,* 651–657.

Kuo, C. K., & Ma, P. X., (2001). Ionically crosslinked alginate hydrogels as scaffolds for tissue engineering: Part 1. Structure, gelation rate and mechanical properties. *Biomaterials, 22,* 511–521.

Kushwah, B. S., & Bhadauria, S., (2010). Development of a biosensor for phenol detection using agarose–guar gum based laccases extracted from *Pleurotusostreatus. J. Appl. Polym. Sci., 115*(3), 1358–1365.

Li, X., Wu, W., Wang, J., & Duan, Y., (2006). The swelling behavior and network parameters of guar gum/poly (acrylic acid) semi-interpenetrating polymer network hydrogels. *Carbohydr. Polym., 66,* 473–479.

Manikoth, R., et al., (2012). Dielectric behavior and pore size distribution of collagen–guar gum composites: Effect of guar gum. *Carbohydr. Polym., 88,* 628–637.

Moreira, L. R. S., & Filho, E. X. F., (2008). An overview of mannan structure and mannan-degrading enzyme systems. *Appl. Microbiol. Biotechnol., 79*(2), 165–178.

Murali, K. P., (2004). Formulation and development of delayed-release tablets of betulinic acid. *Int. J. Pharm. Bio Sci., 5*(1), 512–519.

Murali, R., et al., (2016). Biomimetic hybrid porous scaffolds immobilized with platelet-derived growth factor-BB promote cellularization and vascularization in tissue engineering. *J. Biomed. Mater. Res. A, 104*(2), 388–396.

Murthy, S. N., Hiremath, S. R. R., & Paranjothy, K. L. K., (2004). Evaluation of carboxymethyl guar films for the formulation of transdermal therapeutic systems. *Int. J. Pharm., 272,* 11–18.

Panda, N., et al., (2014). Process optimization, formulation and evaluation of hydrogel (Guar gum-g-poly (acrylamide)) based doxofylline microbeads. *Asian J. Pharm. Clin. Res., 7*(3), 60–65.

Pandey, S., Goswami, G. K., & Nanda, K. K., (2012). Green synthesis of biopolymer–silver nanoparticle nanocomposite: An optical sensor for ammonia detection. *Int. J. Bio. Macromol., 51,* 583–589.

Panigrahy, R. N., Gudipati, S., & Chinnala, K. M., (2016). Design, development and in vitro evaluation of combined floating bioadhesive drug delivery systems of atenolol. *Int. J. Pharm. Pharm. Sci., 8*(2), 41–46.

Prabaharan, M., & Jayakumar, R., (2009). Chitosan-*graft*-*β*-cyclodextrin scaffolds with controlled drug release capability for tissue engineering applications. *Int. J. Bio. Macromol., 44,* 320–325.

Prabaharan, M., (2013). Prospects of biosensors based on chitosan matrices. *J. Chitin Chitosan Sci., 1*(1), 2–12.

Pramanik, N., et al., (2015). Characterization and evaluation of curcumin loaded guar gum/polyhydroxy alkanoates blend films for wound healing applications. *RSC Adv., 5*(78), 63489–63501.

Rafat, M., et al., (2008). PEG-stabilized carbodiimide crosslinked collagen-chitosan hydrogels for corneal tissue engineering. *Biomaterials, 29*, 3960–3972.

Ragothaman, M., Palanisamy, T., & Kalirajan, C., (2014). Collagen-poly(dialdehyde) guar gum based porous 3D scaffolds immobilized with growth factor for tissue engineering applications. *Carbohydr. Polym., 114*, 399–406.

Ramaprasad, Y. V., Krishnaiah, Y. S., & Satyanarayana, S., (1998). In vitro evaluation of guar gum as a carrier for colon-specific drug delivery. *J. Control. Release, 51*, 281–287.

Sathya, S. D., & Prabaharan, M., (2016). Guar gum succinate as a carrier for colon-specific drug delivery. *Int. J. Biol. Macromol., 84*, 10–15.

Sathya, S. D., & Prabaharan, M., (2017). Prabaharan guar gum oleate-graft-poly(methacrylic acid) hydrogel as a colon-specific controlled drug delivery carrier. *Carbohydr. Polym., 158*, 51–57.

Sathya, S. D., et al., (2016). Prabaharan guar gum succinate-sodium alginate beads as a pH-sensitive carrier for colon-specific drug delivery. *Int. J. Biol. Macromol., 91*, 45–50.

Sen, G., et al., (2010). Microwave initiated synthesis of polyacrylamide grafted guar gum (GG-g-PAM) – characterizations and application as matrix for controlled release of 5-aminosalicylic acid. *Int. J. Biol. Macromol., 47*, 164–170.

Sevinc, A. I., et al., (2014). Improvement of colonic healing by preoperative oral partially hydrolyzed guar gum (Bene fiber) in rats which underwent preoperative radiotherapy. *J. Drug Target, 22*, 262–266.

Shahid, M., et al., (2013). Graft polymerization of guar gum with acrylamide irradiated by microwaves for colonic drug delivery. *Int. J. Biol. Macromol., 62*, 172–179.

Shammika, P., Aneesh, T. P., & Viswanad, V., (2017). Formulation and evaluation of synthesized quinazolinone derivative for colon-specific drug delivery. *Asian J. Pharm. Clin. Res., 10*, 207–212.

Sharma, S., et al., (2013). Preparation and characterization of pH-responsive guar gum microspheres. *Int. J. Biol. Macromol., 62*, 636–641.

Sinha, V. R., et al. (2004). Colonic drug delivery of 5-fluorouracil: an in vitro evaluation. *Int. J. Pharm., 269*, 101–108.

Smith, T. J., et al., (2006). A collagen-glycosaminoglycan co-culture model for heart valve tissue engineering applications. *Biomaterials, 27*, 2233–2246.

Soppimath, K. S., Kulkarni, R. A., & Aminabhavi, T. M., (2001). Chemically modified polyacrylamide-g-guar gum-based crosslinked anionic microgels as pH-sensitive drug delivery systems: Preparation and characterization. *J. Control. Release, 75*(3), 331–345.

Soppimath, K. S., Kulkarni, R. A., & Aminabhavi, T. M., (2002). Water transport and drug release study from crosslinked polyacrylamide grafted guar gum hydrogel microspheres for the controlled release application. *Eur. J. Pharm. Biopharm., 53*, 87–98.

Soumya, R. S., Ghosh, S., & Abraham, E. T., (2010). Preparation and characterization of guar gum nanoparticles. *Int. J. Biol. Macromol., 46*, 267–269.

Tembe, S., et al., (2006). Development of electrochemical biosensor based on tyrosinase immobilized in composite biopolymeric film. *Anal. Biochem., 349*, 72–77.

Tembe, S., et al., (2007). Electrochemical biosensor for catechol using agarose–guar gum entrapped tyrosinase. *J. Biotechnol., 128*, 80–85.

Thakur, S., Chauhan, G. S., & Ahn, J. H., (2009). Synthesis of acryloyl guar gum and its hydrogel materials for use in the slow release of L-DOPA and L-tyrosine. *Carbohydr. Polym., 76*, 513–520.

Tiwari, A., & Prabaharan, M., (2009). An amphiphilic nanocarrier based on guar gum-*graft*-poly(ε-caprolactone) for potential drug delivery applications. *J. Biomater. Sci. Polym. Ed.*, *21*, 937–949.

Tiwari, A., et al., (2009). Biodegradable hydrogels based on novel photopolymerizable guar gum-methacrylate macromonomers for in situ fabrication of tissue engineering scaffolds. *Acta Biomaterialia*, *5*, 3441–3452.

Tomolin, J., Taylor, J. S., & Read, N. W., (1989). The effects of mixed fecal bacteria on a selection of various polysaccharides in vitro. *Nutr. Rep. Int.*, *39*, 121–135.

Toti, U. S., & Aminabhavi, T. M., (2004). Modified guar gum matrix tablet for controlled release of diltiazem hydrochloride. *J. Control. Release*, *95*, 567–577.

Tripathy, S., & Das, M. K., (2013). Guar gum: Present status and applications. *Pharm. Scient. Innov.*, *2*(4), 24–28.

Vaghela, C., et al., (2014). Agarose-guar gum assisted synthesis of processable polyaniline composite: Morphology and electro-responsive characteristics. *RSC Adv.*, *4*(104), 59716–59725.

Varma, V. N. S. K., et al., (2016). Development of pH-sensitive nanoparticles for intestinal drug delivery using chemically modified guar gum co-polymer. *Iran. J. Pharm. Res.*, *15*(1), 83–94.

Vinaykumar, K. V., et al., (2011). Colon targeting drug delivery system: A review on recent approaches. *Int. J. Pharm. Biomed. Sci.*, *2*(1), 11–19.

Vivekanandan, K., Gunasekaran, V., & Jayabalan, G., (2015). Design and evaluation of colon-specific drug delivery of budesonide. *Int. J. Pharm. Pharm. Sci.*, *7*, 261–263.

Wong, D., et al., (1997). USP dissolution apparatus III (reciprocating cylinder) for screening of guar-based colonic delivery formulations. *J. Control. Release*, *47*, 173–179.

Wu, J., et al., (2016). Effective symptomatic treatment for severe and intractable pruritus associated with severe burn-induced hypertrophic scars: A prospective, multicenter, controlled trial. *Burns*, *42*(5), 1059–1066.

Zhao, S. P., Ma, D., & Zhang, L. M., (2006). New semi-interpenetrating network hydrogels: Synthesis, characterization and properties. *Macromol. Biosci.*, *6*, 445–451.

CHAPTER 6

Pharmaceutical Applications of Locust Bean Gum

MD SAQUIB HASNAIN,[1] AMIT KUMAR NAYAK,[2]
MOHAMMAD TAHIR ANSARI,[3] and DILIPKUMAR PAL[4]

[1]*Department of Pharmacy, Shri Venkateshwara University, NH-24, Rajabpur, Gajraula, Amroha – 244236, U.P., India*

[2]*Department of Pharmaceutics, Seemanta Institute of Pharmaceutical Sciences, Mayurbhanj – 757086, Odisha, India*

[3]*Department of Pharmaceutical Technology, Faculty of Pharmacy and Health Sciences, Universiti Kuala Lumpur Royal College of Medicine Perak, Malaysia*

[4]*Department of Pharmaceutical Sciences, Guru Ghasidas Vishwavidyalaya, Koni, Bilaspur – 495009, C.G., India*

6.1 INTRODUCTION

An inspiring array of various materials, provided by nature, have been finding very attractive applications in drug delivery as well as biomedical applications (Hasnain et al., 2010, 2019; Hati et al., 2014; Jana et al., 2013; Jena et al., 2012a,b; Sinha Mahapatra et al., 2011). Nowadays, there is a trend to mimic the nature, and for this purpose, there are no other better candidates than the proper materials obtained from nature for such a task (Hasnain et al., 2010; Nayak and Pal, 2012; Nayak et al., 2013). Amongst the various naturally derived materials, naturally occurring biopolymers demonstrate as an extraordinary example, how all the properties exhibited by the biological materials and systems are absolutely determined by the monomers and their sequences physico-chemical properties (Malafaya et al., 2007; Nayak and Pal, 2015). These naturally occurring materials have a number of outstanding merits over the synthetic materials, specifically improved capability for the adhesion of cells and mechanical properties analogous to the natural tissues (Hasnain et al., 2019; Malafaya et al., 2007). In addition, these naturally occurring materials are very economical, easily available, non-toxic in nature, typically biodegradable and biocompatible with few exceptions (Hasnain and Nayak, 2018a,b; Pal et al., 2018). In contrast, some inherent constraints are also to take into consideration, such as the utmost possibility of immunogenicity and variability in polymer associated to both, i.e., origin as well as supplier level (Pal and Nayak, 2015a,b).

These naturally occurring polysaccharides are the biological macromolecules comprising of one or more monosaccharides linked together in a long sequence (Nayak et al., 2018a,c). These monosaccharides are linked with each other in different ways to present the linear or branched polysaccharides. These natural polysaccharides are the products of photosynthesis, which is a natural carbon-capture process, followed by the additional biosynthetic alterations. The rising interest in polysaccharides nowadays relies on the invention of fresh synthetic routes for the chemical modifications of these polysaccharides, which allow the new biological activities promotion and the modifications of their properties for explicit purposes (Laurienzo, 2010; Nayak and Pal, 2018a). In last few decades, natural polysaccharides have been unearthing very attractive and useful applications in drug delivery as well as biomedical applications (Hasnain and Nayak, 2018b; Malakar and Nayak, 2013; Malakar et al., 2012, 2014; Nayak et al., 2011, 2018; Ray et al., 2018). In general, these naturally obtained polysaccharides nowadays playing lead roles as the thickening agent, suspending agents, emulsifying agent, hydrating agent, as well as gelling agent and results in different kind of applications in aforementioned disciplines (Malafaya et al., 2007; Nayak et al., 2014a, b, c, d, 2018). Natural polysaccharides commonly occur in various plants, microorganisms, algae, as well as in animals (Pal and Nayak, 2015a,b). However, the majority of natural polysaccharides have been isolated or extracted from various plant resources (Nayak and Pal, 2016a; Nayak et al., 2015; Pal and Nayak, 2017). Some important and widely used/investigated plant polysaccharides are gum Arabica (GA) (Nayak et al., 2012a), guar gum (Soumya et al., 2010), pectin (Nayak and Pal, 2013c; Nayak et al., 2013a, 2014b,c), sterculia gum (Bera et al., 2015a,b; Guru et al., 2013; Nayak and Pal, 2016b), tamarind gum (TmG) (Nayak, 2016; Nayak and Pal, 2013b, 2017b; Nayak et al., 2013e, 2014e, 2016), locust bean gum (LBG) (Dionísio and Grenha, 2012), cashew gum (Das et al., 2013, 2014; Hasnain et al., 2017a,b, 2018a), dillenia gum (Ketousetuo and Bandyopadhyay, 2007), okra gum (Sinha et al., 2015a,b), fenugreek polysaccharide (Nayak and Pal, 2014; Nayak et al., 2013b,f), linseed polysaccharide (Hasnain et al., 2018b), gum odina (Jena et al., 2018), starches (Malakar et al., 2013; Nayak and Pal, 2013a, 2017a; Nayak et al., 2013c, 2014a), ispaghula mucilage (Guru et al., 2018; Nayak et al., 2010a, 2013d, 2014d), etc. These plant-derived polysaccharides are isolated or extracted from various plant parts like seeds, fruits, leaves, pods, exudates, roots, etc. (Nayak et al., 2010c, 2018b; Pal and Nayak, 2017; Pal et al., 2010). However, a huge volume of plant polysaccharides obtained from the seed endosperms of various plant species (Nayak et al., 2012b). The gums or polysaccharides are obtained from the endosperm of a range of plants seeds (chiefly from Leguminosae), where these functioned as the reserve materials, which is utilized during germination (Silveira and Bresolin, 2011). The majority of these plant-derived natural polysaccharides have basic similar structural moieties, which are known as galactomannans (Dionísio and Grenha, 2012; Silveira and Bresolin, 2011). Hence, the galactomannans are the natural polysaccharides, which mainly consists of the monosaccharide, mannose, and galactose units. The mannose part present in the linear chain linked as the side chain with galactopyranosyl residues at varying distances depending on the origin of the plant (Silveira and Bresolin, 2011). Resembling other galactomannans, LBGis also obtained from the endosperm of *Ceratonia siliqua* Linn., seeds which belong to the *Fabaceae* family (Dionísio and Grenha, 2012; Prajapati et al., 2013). The best applications of LBG areas, the thickening and stabilizing agent in both food and cosmetic industries (Pollard et al., 2007; Sujja-areevath et al., 1998). LBG has been recording augmented importance in various pharmaceutical and biomedical applications. Hence, this chapter recounts with the constructive and inclusive discussion on the pharmaceutical applications of LBG. Furthermore, several important features of LBG, i.e., source, extraction, chemical composition, and characteristics are also described in brief.

6.2 SOURCES AND EXTRACTION

LBG is obtained from the seeds of *Ceratonia siliqua* (carob tree), which is incredibly available in the Mediterranean region. However, its geographical distribution of LBG expands to different regions of Asia, South America, and North Africa. This plant polysaccharide also possesses several other synonyms reported in the literature like carob flour, carob bean gum, carob seed gum, or even *Ceratonia* (Rowe et al., 2006).

Ceratonia siliqua seeds represent roughly 10% of the weight of *Ceratonia siliqua* fruit processed industrially by hull cracking, sifting, followed by milling operations to isolate and grind the endosperms, and then sold as crude flour (Bouzouita et al., 2007; Dakia et al., 2008). These seeds of *Ceratonia siliqua* principally contain galactomannan, which covers around 80%, the rest 20% are proteins and impurities (Andrade et al., 1999). In LBG, the protein content was reported, which comprises around 32% albumin and globulin, whereas the rest 68% correspond to glutelin (Smith et al., 2010). Ash and insoluble acid materials are the main impurities (Bouzouita et al., 2007). After the processing of seeds, raw galactomannan may further deposited for a number of processes like enzymatic or alkaline hydrolysis, precipitation by ethanol or isopropanol, and purification with the help of methanol, or by complexes of copper or barium to remove the protein content as well as the impurities (Andrade et al., 1999; Bresolin et al., 1999). Generally, these impurities still remain insoluble when heating at a temperature of up to 70°C (Kök et al., 1999). For the removal of proteins, precipitation by means of isopropanol found to be pretty efficient. The general flow chart of LBG extraction is presented in Figure 6.1.

The kernels of carob are intricate to process, as the coat of seeds is very sturdy and hard. The kernels are peeled by means of special processes to avoid the damaging of endosperm and the germ. Acid peeling process and thermal peeling process are usually applied to extract LBG in larger volume (Prajapati et al., 2013).

1. **Acid Peeling Process:** The kernels of carob are treated with sulfuric acid at some fix temperature to carbonize the seed coat. The rest of the seed coat fragments are aloofedon or after the clean endosperm in an effective washing and brushing process. These peeled carob kernels are subsequently dried, cracked, and the more friable germ gets crushed. From the complete or unbroken endosperm halves, the germ parts are separated off. LBG obtained from this procedure is whitish and has higher viscosity (Prajapati et al., 2013).

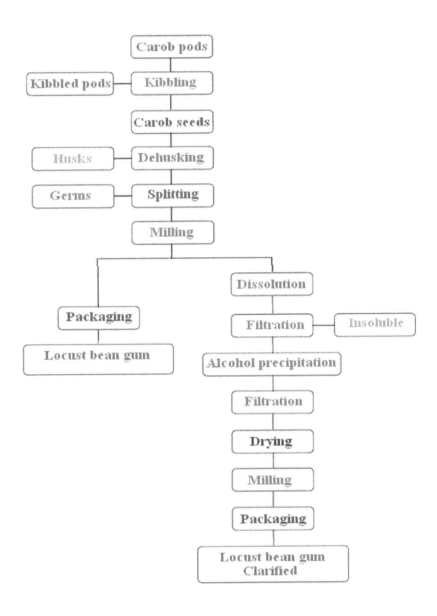

General flow chart of LBG extraction

FIGURE 6.1 **(See color insert.)** General flow chart of LBG extraction.

2. **Thermal Peeling Process:** The carob kernels might be heated in a rotating furnace where the coats of the seeds relatively pop off from the remaining. The endosperm halves are recuperated from crashed germs as well as a burned husk. This procedure yields a product of comparatively darker color (Prajapati et al., 2013). The consequence is that no sulfuric acid required as the processing aid and as a result, no effluent instigates from this method of production. Subsequently, the isolated endosperm is pulverized to a fine powder which stands for the concluding product of LBG (Prajapati et al., 2013).

6.3 CHEMICAL COMPOSITION AND CHARACTERISTICS

1. LBG primarily consists of a neutral polymer known as galactomannan, which is made up of 1,4-linked units of d-mannopyranosyl and every fourth chain in five is substituted on C6 with a unit of d-galactopyranosyl (Deaand Morrison, 1975; Venkataraju et al., 2007). This is to believe that the d-galactose to d-mannose ratio differs due to the different origins of these gum materials and plants growth conditions during production. LBG is made up of a high molecular weight polysaccharide which is composed of galactomannans units consisting (1–4)-linked β-d-mannose skeleton by means of (1–6)-linked α-d-galactose side chains (Daas et al., 2002; Mathur and Mathur, 2005). Figure 6.2 presents the chemical structure of LBG, consisting of mannose and galactose.

2. LBGhas the ability to form exceptionally viscous solutions at comparatively low concentrations, which are more or less unaltered by pH, salts, or temperature (Alves et al., 1999; Mathur and Mathur, 2005). In reality, as a neutral polymer, their viscosity, as well as solubility, is slightly exaggerated by the changes in pH within the range of 3–11 (Dionísio and Grenha, 2012). LBG is biocompatible, bioresorbable, and biodegradable whereas according to Joint Expert Committee of FAO/WHO on Food Additives which was held in Geneva on April' 75, these LBG are non-teratogenic and non-mutagenic (Prajapati et al., 2013).

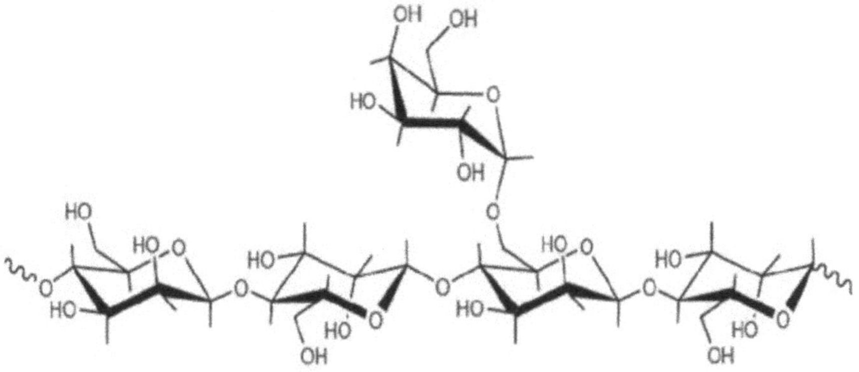

FIGURE 6.2 Chemical structure of LBG.

3. The *in vivo* biodegradation of LBG is likely to be driven by enzymatic activity normally as the numerous enzymes present in the organism might cleave the LBG macromolecule. Oral route of administration of polysaccharide is possibly the most suitable route, which effortlessly ensures an effectual degradation of these polysaccharides. This happens due to the β-mannanase occurrence in the colonic region of human and accordingly, this occurrence serves as the base for the growth of several strategies of colonic drug delivery systems (Alonso-Sande et al., 2009; Nakajima et al., 1997). The absolute LBG degradation entails the other two enzymes: β-mannosidase and α-galactosidase. These enzymes act on mannose and galactose moieties correspondingly, which produced d-mannose and d-galactose. The abovementioned enzymes were also observed in human fecal contents, strengthening the potential of LBG applications in colonic delivery, although it also ensures their degradation upon any oral administration modality (Jain et al., 2007). In fact, there are some reports available about LBG *in-vivo* degradation mediated by the bacteria present in the colon, a consequence that has been explicitly accredited to *Bacteroides* and *Ruminococcus* (Bauer and Kesselhut, 1995; Jain et al., 2007).

4. LBG, in 1983, was the first time reported for having a hypolipidemic activity that lowers LDL (low-density lipoprotein) cholesterol because of the higher insoluble fiber contents (Zavoral et

al., 1983). A number of successive researches demonstrated the positive effects of LBG for hypercholesterolemia control (Ruiz-Rosoet al., 2010; Zunft et al., 2003). It has also been suggested for diabetes treatment (Brennan, 2005; Dionísio and Grenha, 2012). Additionally, attributable to the high gelling ability and the fact that the formed gel is not assimilated by the gastro-intestinal tract (GIT). Ingestion causes a sensation of satiety, resulting in decreased absorption of nutrients, thus rendering LBG adequate for inclusion in dietary products (Dakia et al., 2007; Urdiain et al., 2004).

5. With other polysaccharides, LBG displays a substantial ability to form synergistic interactions primarily because of the frequent-OH groups available in polysaccharides molecular structure (Sandolo et al., 2010; Srivastava and Kapoor, 2005). The flexibility of the polymer augmented by this synergy and, in several cases, allows the development of gelling structures having significant biopharmaceutical applications. Furthermore, very important synergies of LBG are observed with carrageenan and xanthan gum (Fernandes et al., 1991; Lundin and Hermansson, 1995). The details of these interactions will be described in a later section. Some reports also suggested LBG and guar gum interactions even though these barely result in augmented viscosity of the solution and do not help in the true gel formation (Dionísio and Grenha, 2012).

6.4   PHARMACEUTICAL APPLICATIONS

6.4.1   APPLICATIONS IN ORAL DRUG DELIVERY

Oral route of drug administration has been the most suitable and universally employed in drug delivery among the different routes of drug administration (Behera et al., 2010; Nayak et al., 2010b; Nayak and Malakar, 2010, 2011; Nayak and Manna, 2011). LBG applications in orally administered drug delivery systems are primarily focused on its application in tablets as the matrix forming materials, advantaging from the actuality that polysaccharides are normally considered to have a significant role in matrixes drug release mechanisms. LBG in these systems typically projected to afford systemic drug absorption and contributes with its swelling ability to offer a controlled release of the drug. Additionally, in the majority of the cases, it is noticed that LBG involvement with the second polymer offers an enhanced effect, advantaging from explicit interactions taking place between the polymers.

In 1998, Sujja-areevath et al., reported the first work of LBG application as the single polysaccharide excipients in tablet formulations. They formulated the sodium diclofenac mini-matrixes, which contained 49.5% LBG. These optimized tablet formulations had a 1:1ratio of drug/polymer as the high ratios led to the matrix integrity loss. LBG containing preparations indicated lower swelling (50% in 6 h) as compared to those having xanthan (250% in 6 h) or karaya gum (150% in 6 h) and was reported to follow more or less Fick's law of diffusion for the swelling rate. Conversely, release kinetics of drug and polymeric erosion were found non-Fickian and, in comparison of other gums, LBG based tablets exhibited the fastest erosion at pH 7.0 of phosphate buffer solution (65% versus 45% and 25% for xanthan and karayagum, respectively).

In 2007, Coviello et al., and their team designed a matrix tablet of LBG containing theophylline and myoglobin. With glutaraldehyde (GA), this tablet was cross-linked further in an effort to offer a network with the potential for an effectual controlled release. Although, that effect was not experienced as the release rate is identical to both, i.e., in attendance and nonattendance of the cross-linker (80% in 8 h). Justification of such observation is based on the fact that LBG has just a few side chains and barely a decreased number of cross-linking takes place within the polymeric network, which is not sufficient to influence the mechanism of drug release. On the contrary, guar gum containing tablets has a maximum number of side chains and consequently permits stronger cross-linking, confirm a large variation in drug release profile among the matrixes of cross-linked and non-cross-linked. Two exceptionally latest researches showed the capability of LBG to work as a super disintegrant in orodispersible tablets in a different approach but still in the ambit of employing LBG in the formulations as a single polysaccharide (Malik et al., 2011a). One of the formulations of nimesulide tablets, in which 10% of LBG inclusion occurred in a disintegration time of 13seconds. If cross-carmellose sodium, a standard super disintegrant in place of LBG was employed, this time was doubled. The other pharmaceutical product made of ofloxacin-loaded Eudragit® microsphere was employed to formulate the orodispersible tablets by using LBG (Malik et al., 2011b). These polysac-charides were employed in a varying concentration within the range of 2.5 and 10% and the disintegration time was found to be varied in between 12 and 20 seconds, which found to be decreased as the LBG concentration increased.

For the design of systems, this is a very common approach to combine LBG with other polysaccharides, in numerous cases advantaging from material synergies, as described in the earlier example. Tablets from LBG/xanthan gum hydrogels containing myoglobin were prepared by compressing these freeze-dried hydrogels (Sandolo et al., 2007). Primarily LBG governed the release behavior, which slows down the diffusion of the drug from the matrix. Even in a ratio of 1/9 LBG/xanthan gum was when employed, the drug released was only 44% in 24 h in water. In a different research, it was observed that LBG by itself does not control the diltiazem hydrochloride release from tablets, however on the addition of karaya gum to the matrix showed the controlled release effect (LBG/karaya gum ratio of 1/1) (Moin and Shivakumar, 2010). Although, this result required the attendance of the polymeric mixture and the drug in a 1:2 concentration (drug/polymer ratio = 1/2).

In multi-layered matrix tablets development, another different approach applied which also give modified release behavior. These dosage forms are drug delivery devices comprising of a matrix core containing the active solute and one or more modulating layers, which are incorporated during the tableting process. Controlling the hydration rate of matrix core is facilitated by the modulating layers, thus confining the existing surface area for drug diffusion and at the meantime controlling the rate of solvent penetration (Colombo et al., 1990). On this basis, matrix tablets of sodium diclofenac with LBG, xanthan gum, or in a ratio of 1:1 of both polymers were prepared (Ahmed et al., 2010). More than 90% of the drug released from these matrix tablets in 12 h whereas the LBG containing system indicated the fastest release. A carboxymethyl cellulose triple external layer performed as the release retardant for hydrophilic matrix core of controlled release of diclofenac and was almost 70% for core tablets of LBG in the equal time.

To enhance the solubility of the drug, solid dispersions containing LBG and lovastatin were prepared (Patel et al., 2008). To reduce the viscosity, LBG was earlier submitted to thermal treatment and was observed that this did not affect its swelling ability. This action had a comprehensible consequence on the solubility of lovastatin by itself, which was studied in pH 7 for 2 h. The release of lovastatin was more or less 53% from pure LBG solid dispersions preparation, whereas 65% was from treated LBG. Merely, 35% of drugs released from the unformulated lovastatin in equal time, by taking into account, this solubility enhancement is significant. For the development of solid dispersion, several other techniques were also assessed. It was found that the modified solvent evaporation method showed superior results followed by spray-drying. In comparison to the unformulated lovastatin, the formulated solid dispersions tested *in vivo* also showed improved therapeutic results (Patel et al., 2008). Hydrogels with LBG and xanthan gum were prepared in different research to control the prednisolone release (Watanabe et al., 1992). The rate of drug release from the hydrogels reduced as the concentration of gum augmented, which indicated that diffusion of the drug was primarily controlled by the density of the three-dimensional matrix structure. As the additives in the hydrogel, glycerin, and sucrose were employed, this demonstrated a major decline in the release rate of the drug.

Another approach, namely microspheres, is found moderately different and concerns the application of multiparticulate systems. The microspheres of LBG/alginate (Deshmukh et al., 2009a) and LBG-xanthan gum/alginate (Deshmukh et al., 2009b) containing diclofenac sodium (DS) were formulated by employing ionic cross-linking method by means of calcium ions. It was observed that drug association efficiency was more than 90% and reported that as the quantities of LBG or the mixture of gums increases, it increases. It was observed that, in simulated gastric (pH 1.2; 2 h) and intestinal (pH 7.2; 10 h) fluids, both the microspheres exhibited a profile of controlled release for 12 h. Both of these mentioned articles inadequately discussed and had very limited details, unfortunately. Hence, it was unfeasible to conclude on the basis of these for more appropriate formulation diclofenac delivery and their release.

6.4.2 APPLICATIONS IN BUCCAL DRUG DELIVERY

There are two most important advantages of drug administration through the buccal mucosa, which comprise avoiding pre-systemic elimination within the GIT and first-pass hepatic effect (Mujoriya et al., 2011). Thus, buccal drug delivery primarily envisages improving poorly absorbable drugs bioavailability in the intestinal region (Şenel, 2010). Strong muco-adhesiveness is one of the main characteristics to be revealed by buccal delivery systems, which is typically obtained by means of mucoadhesive polymers (Ratha et al., 2010). It was observed that LBG as compared to other polysaccharides like chitosan (CS) has a less stronger mucoadhesive profile (Yamagar et al., 2010).

In order to enhance the metoprolol bioavailability, buccoadhesive tablets of LBG or LBG and xanthan gum mixtures as matrix materials were formulated by avoiding an extensive first-pass effect of the drug (Yamagar et al., 2010). Only LBG containing preparations with 7.5% of polymer produced a progressive release of the drug, which led to release of 98% of the drug in 45 min, whereas an increment in LBG up to 15% resulted in reduced release rate, offering just about 45% of drug release in equal time. Taking into account of physical integrity, hardness, and mucoadhesive strength, xanthan gum, and LBG mixture showed more effective for tablet preparations. Those tablets with 2:1 ratio of LBG/xanthan gum demonstrated whole drug release as desired in 45 min, but also have abridged permeation of drug. To triumph over this inadequacy, incorporation of 1% sodium lauryl sulfate was done in the preparation, which resulted in enhanced drug permeation across the buccal mucosa of porcine. Vijayaraghavan et al. (2008), in another study, also employed LBG in a combination of a different polymer for the development of LBG/CS tablets containing propranolol hydrochloride in different ratios (2/3, 3/2 and 4/1) and administered to human volunteers for comparing the bioavailability of propranolol hydrochloride. In a comparison to a similar preparation for oral administration, all buccal preparations have enhanced the bioavailability of drug (1.3, 2.1 and 2.3 folds, correspondingly). Naturally, superior mucoadhesion were demonstrated by those tablets which contains polymer in a ratio of 2/3, due to utmost content of CS, and were also indicating more comprehensive release (98% in 10 h, as compared to 92% and 90% of formulations 3/2 and 4/1, respectively) even though variations are not incredibly significant.

6.4.3   APPLICATIONS IN COLONIC DRUG DELIVERY

The rationale behind the application of polysaccharides in the development of drug delivery systems, which is intended to the colonic delivery of drugs, primarily relies on the attendance of huge quantities of the polysaccharides in the colon of human. This is an outcome of the truth that numerous predominantly colonized bacteria are present in this region, which produces a number of enzymes (Jain et al., 2007). Regardless of the evident use for local therapeutic action, for example in colonic inflammatory diseases, systemic delivery of drugs for the colon is also an alternative, in particular for those drugs noticing complex absorption from the upper GIT. This possibility originates from the truth that a variety of digestive enzymes are present in the upper regions of GIT which lacks in the colon, primarily proteinases, consequently enjoying a less intimidating milieu as compared to the stomach and small intestine (Chourasia and Jain, 2004; Sinha and Kumria, 2001). It is well known that in the colon of human, β-mannanase, and other pertinent enzymes are present which ensures LBG *in vivo* degradation (Jain et al., 2007; Sinha and Kumria, 2001). It was observed that *Bacteroides* and *Ruminococcus* bacterial species were involved in the degradation of LBG (Chourasia and Jain, 2004; Kumar et al., 2009).

Another research based on the application of LBG in colonic drug delivery consists of the preparation of butanediol diglycidyl ether cross-linked LBG films, which were employed as a coating material in theophylline tablets (Hirsch et al., 1999). These films indicated a quite high swelling capability (300–500%) and were revealed to endure degradation by colonic microflora, which potentiating it in colonic drug delivery application. However, it was found that it was not suitable for application in colonic carrier preparation as the mechanical instability of the films was reported, particularly at higher coating amounts. A different research showed the LBG/CS mixtures probability as the coating materials to offer the safeguard from the stomach and small intestine physiological environment, while allowing colonic bacterial enzymes degradation, facilitating the release of the drug. Several different ratios of LBG/CS were investigated as the coating material over core tablets of mesalazine. The capability of LBG to hydrate and production of a layer of viscous gel were anticipated to offer a slower dissolution towards the core tablet. The 4:1 ratio exhibited the most satisfactory behavior illustrating 98% of cumulative release after the incubation of 26 h, which was in correspondence to 2 h in HCl, 3 h in pH 7.4 buffer and 21 h in Phosphate buffer solution of pH 6.8 having 4% (w/v) contents of rat cecal. Human *in vivo* investigations demonstrated that release of the drug commenced only after 5 h, which corresponds to the small intestine's transit time (Kumar et al., 2009).

6.4.4   *APPLICATIONS IN OCULAR DRUG DELIVERY*

Another distinctive study suggested the application of LBG in the ocular drug delivery system. Microparticles of the LBG/*i*-carrageenan which encapsulating gentamicin was formulated by means of emulsification and further to be included in a gel of poly (vinyl alcohol) (PVA) that is employed on the surface of the eye. An initial burst release was observed within the first 6 h in the formulations with no LBG, which by the addition of 10% LBG declined by over 50% (Suzuki and Lim, 1994). Regrettably, no other research was reported on this system for its potential to allow a broader vision.

6.4.5   *APPLICATIONS IN TOPICAL DRUG DELIVERY*

In preparation of topical formulation, LBG was used and discussed. In this, researchers formulated LBG/xanthan gum containing hydrogel, which contains these gums in a ratio of 1/1, and was employed to include in niosomes (Marianecci et al., 2011). These niosomes, which are non-ionic surfactant vesicles, tender a number of advantages above traditional liposomes including lesser costs, superior chemical stability, and materials maximum availability (Lilia Romero and Morilla, 2011; Sinico and Fadda, 2009). The niosomes were prepared by loading a number of different drugs like calcein, ibuprofen, and caffeine. The consequent inclusion of niosomes on the hydrogel offered a protecting effect on vesicle integrity and a sluggish drug release up to 50 h from the polysaccharide system (Marianecci et al., 2011).

6.5   CONCLUSION

Basically assisting the manufacturing progression, excipients have been incorporated in the formulations of drugs as inert materials conventionally. Yet, they have been more and more incorporated in dosage forms to execute specific functions intended at improving drug delivery in recent decades. LBG is being employed in more than a few different functions, ranging from excipients for controlled release to disintegrant of tablet in biopharmaceutical applications. In the majority of the cases, a second substance is coupled with the polysaccharide which benefiting from strong synergies, and fields of application are varied, including oral, buccal, and colonic delivery, but applications in ocular and topical drug applications have been discussed.

6.6 SUMMARY

LBG is obtained from the seeds of the *Ceratonia siliqua* (carob tree) and belongs to the polysaccharide group of galactomannans, which is made of mannose and galactose units, and hence, belongs to the galactomannans category. It is a neutral polysaccharide, which is frequently employed in the development of oral delivery systems, while a number of works also report its pharmaceutical application in topical, ocular, buccal as well as in colonic drug delivery systems. This chapter messes with an inclusive and effective discussion on pharmaceutical uses of LBG with some of the significant characteristics of LBG like source, extraction, chemical composition, and their application in pharmacy.

KEYWORDS

- **drug delivery**
- **locust bean gum**
- **pharmaceutical excipient**
- **polysaccharide**

REFERENCES

Ahmed, S., Mangamoori, L., & Rao, Y., (2010). Formulation and characterization of matrix and triple-layer matrix tablets for oral controlled drug delivery. *Int. J. Pharm. Pharmaceut. Sci., 2*, 137–143.

Alonso-Sande, M., Teijeiro-Osorio, D., Remuñán-López, C., & Alonso, M. J., (2009). Glucomannan, a promising polysaccharide for biopharmaceutical purposes. *Eur. J. Pharm. Biopharm., 72*, 453–462.

Alves, M. M., Antonov, Y. A., & Gonçalves, M. P., (1999). The effect of structural features of gelatin on its thermodynamic compatibility with locust bean gum in aqueous media. *Food Hydrocol., 13*, 157–166.

Andrade, C., Azero, E., Luciano, L., & Gonçalves, M., (1999). Solution properties of the galactomannans extracted from the seeds of Caesalpiniapulcherrima and Cassia javanica: Comparison with locust bean gum. *Int. J. Biol. Macromol., 26,* 181–185.

Bauer, A., & Kesselhut, A., (1995). Novel pharmaceutical excipients for colon targeting. *STP Pharm. Sci., 5,* 54–59.

Behera, A. K., Nayak, A. K., Mohanty, B., & Barik, B. B., (2010). Formulation and optimization of losartan potassium tablets. *Int. J. Appl. Pharm., 2,* 15–19.

Bera, H., BodduPalli, S., & Nayak, A. K., (2015b). Mucoadhesive-floating zinc-pectinate-sterculia gum interpenetrating polymer network beads encapsulating ziprasidone HCl. *Carbohydr. Polym., 131,* 108–118.

Bera, H., Kandukuri, S. G., Nayak, A. K., & BodduPalli, S., (2015a). Alginate-sterculia gum gel-coated oil-entrapped alginate beads for gastroretentive risperidone delivery. *Carbohydr. Polym., 120,* 74–84.

Bouzouita, N., Khaldi, A., Zgoulli, S., Chebil, L., Chekki, R., & Chaabouni, M., (2007). The analysis of crude and purified locust bean gum: A comparison of samples from different carob tree populations in Tunisia. *Food Chem., 101,* 1508–1515.

Brennan, C. S., (2005). Dietary fiber, glycaemic response, and diabetes. *Mol. Nutr. Food Res., 49,* 560–570.

Bresolin, T. M., Milas, M., Rinaudo, M., Reicher, F., & Ganter, J. L., (1999). Role of galactomannan composition on the binary gel formation with xanthan. *Int. J. Biol. Macromol., 26,* 225–231.

Chourasia, M., & Jain, S., (2004). Polysaccharides for colon targeted drug delivery. *Drug Deliv., 11,* 129–148.

Colombo, P., Conte, U., Gazzaniga, A., Maggi, L., Sangalli, M. E., & Peppas, N. A., (1990). Drug release modulation by physical restrictions of matrix swelling. *Int. J. Pharm., 63,* pp. 43–48.

Coviello, T., Alhaique, F., Dorigo, A., Matricardi, P., & Grassi, M., (2007). Two galactomannans and scleroglucan as matrices for drug delivery: Preparation and release studies. *Eur. J. Pharm. Biopharm., 66,* 200–209.

Daas, P., Grolle, K., Vliet, T., Schols, H., & De, J. H., (2002). Toward the recognition of structure-function relationships in galactomannans. *J. Agr. Food Chem., 50,* 4282–4289.

Dakia, P. A., Wathelet, B., & Paquot, M., (2007). Isolation and chemical evaluation of carob (*Ceratonia siliqua* L.) seed germ. *Food Chem., 102,* 1368–1374.

Dakia, P., Blecker, C., Robert, C., Whatelet, B., & Paquot, M., (2008). Composition and physicochemical properties of locust bean gum extracted from whole seeds by acid or water dehulling pre-treatment. *Food Hydrocol., 22,* 807–818.

Das, B., Dutta, S., Nayak, A. K., & Nanda, U., (2014). Zinc alginate-carboxymethyl cashew gum microbeads for prolonged drug release: Development and optimization. *Int. J. Biol. Macromol., 70,* 505–515.

Das, B., Nayak, A. K., & Nanda, U., (2013). Topical gels of lidocaine HCl using cashew gum and Carbopol 940: Preparation and in vitro skin permeation. *Int. J. Biol. Macromol., 62,* 514–517.

Dea, I. C., & Morrison, A., (1975). Chemistry and interactions of seed galactomannans. *Adv. Carbohydr. Chem. Biochem., 31,* 242–312.

Deshmukh, V., Jadhav, J., Masirkar, V., & Sakarkar, D., (2009b). Formulation, optimization and evaluation of controlled release alginate microspheres using synergy gum blends. *Res. J. Pharm. Technol., 2,* 324–327.

Deshmukh, V., Sakarkar, D., & Wakade, R., (2009a). Formulation and evaluation of controlled release alginate microspheres using locust bean gum. *J. Pharm. Res., 2,* 458–461.

Dionísio, M., & Grenha, A., (2012). Locust bean gum: Exploring its potential for biopharmaceutical applications. *J. Pharm. Biol. Sci., 4,* 175–185.

Fernandes, P. B., Gonçalves, M. P., & Doublier, J. L., (1991). A rheological characterization of kappa-carrageenan/galactomannan mixed gels: A comparison of locust bean gum samples. *Carbohydr. Polym., 16,* 253–274.

Guru, P. R., Bera, H., Das, M., Hasnain, M. S., & Nayak, A. K., (2018). Aceclofenac-loaded Plantago ovata F. husk mucilage-Zn^{+2}-pectinate controlled-release matrices. *Starch –Stärke, 70,* 1700136.

Guru, P. R., Nayak, A. K., & Sahu, R. K., (2013). Oil-entrapped sterculia gum-alginate buoyant systems of aceclofenac: Development and in vitro evaluation. *Colloids Surf. B.: Biointerf., 104,* 268–275.

Hasnain, M. S., & Nayak, A. K., (2018a). Alginate-inorganic composite particles as sustained drug delivery matrices, In: Inamuddin, A. M., & Asiri, A. M., (eds.), *Applications of Nanocomposite Materials in Drug Delivery* (pp. 39–74). A volume in Woodhead Publishing Series in Biomaterials, Elsevier Inc.

Hasnain, M. S., & Nayak, A. K., (2018b). Chitosan as a responsive polymer for drug delivery applications, In: *Makhlouf, A. S. H., & Abu-Thabit, N. Y., (eds.), Stimuli-Responsive Polymeric Nanocarriers for Drug Delivery Applications* (Vol. 1, pp. 581–605). Types and triggers, Woodhead Publishing Series in Biomaterials, Elsevier Ltd.

Hasnain, M. S., Ahmad, S. A., Chaudhary, N., Hoda, M. N., & Nayak, A. K., (2019). Biodegradable polymer matrix nanocomposites for bone tissue engineering, In: Inamuddin, A. M., & Asiri, A. M., (eds.), *Applications of Nanocomposite Materials in Orthopedics* (pp. 1–37). A volume in Woodhead Publishing Series in Biomaterials, Elsevier Inc.

Hasnain, M. S., Nayak, A. K., Singh, R., & Ahmad, F., (2010). Emerging trends of natural-based polymeric systems for drug delivery in tissue engineering applications. *Sci. J. UBU., 1,* 1–13.

Hasnain, M. S., Rishishwar, P., & Ali, S., (2017a). Use of cashew bark exudate gum in the preparation of 4% lidocaine HCL topical gels. *Int. J. Pharm. Pharmaceut. Sci., 9,* 146–150.

Hasnain, M. S., Rishishwar, P., & Ali, S., (2017b). Floating-bioadhesive matrix tablets of hydralazine HCL made of cashew gum and HPMC K4M. *Int. J. Pharm. Pharmaceut. Sci., 9,* 124–129.

Hasnain, M. S., Rishishwar, P., Rishishwar, S., Ali, S., & Nayak, A. K., (2018a). Extraction and characterization of cashew tree (*Anacardium occidentale*) gum, use in aceclofenac dental pastes. *Int. J. Biol. Macromol., 116,* 1074–1081.

Hasnain, M. S., Rishishwar, P., Rishishwar, S., Ali, S., & Nayak, A. K., (2018b). Isolation and characterization of linumusitatisimum polysaccharide to prepare mucoadhesive beads of diclofenac sodium. *Int. J. Biol. Macromol., 116,* 162–172.

Hati, M., Jena, B. K., Kar, S., & Nayak, A. K., (2014). Evaluation of anti-inflammatory and antipyretic activity of *Carissa carandas*, L. leaf extract in rats. *Int. J. Pharm., 1,* 18–25.

Hirsch, S., Binder, V., Schehlmann, V., Kolter, K., & Bauer, K. H., (1999). Lauroyldextran and crosslinked galactomannan as coating materials for site-specific drug delivery to the colon. *Eur. J. Pharm. Biopharm., 47,* 61–71.

Jain, A., Gupta, Y., & Jain, S., (2007). Perspectives of biodegradable natural polysaccharides for site-specific drug delivery to the colon. *J. Pharm. Pharmaceut. Sci., 10,* 86–128.

Jana, S., Das, A., Nayak, A. K., Sen, K. K., & Basu, S. K., (2013). Aceclofenac-loaded unsaturated esterified alginate/gellan gum microspheres: *In vitro* and *in vivo* assessment. *Int. J. Biol. Macromol., 57,* 129–137.

Jena, A. K., Nayak, A. K., De, A., Mitra, D., & Samanta, A., (2018). Development of lamivudine containing multiple emulsions stabilized by gum odina. *Future J. Pharm. Sci. 4,* 71–79.

Jena, B. K., Ratha, B., Kar, S., Mohanta, S., & Nayak, A. K., (2012b). Antibacterial activity of the ethanol extract of *Ziziphusxylopyrus* Willd.(Rhamnaceae). *Int. J. Pharma Res. Rev., 1,* 46–50.

Jena, B. K., Ratha, B., Kar, S., Mohanta, S., Tripathy, M., & Nayak, A. K., (2012a). Wound healing potential of Ziziphusxylopyrus Willd. (Rhamnaceae) stem bark ethanol extract using *in vitro* and *in vivo* model. *J. Drug Deliv. Ther., 2,* 41–46.

Ketousetuo, K., & Bandyopadhyay, A. K., (2007). Development of oxytocin nasal gel using natural mucoadhesive agent obtained from the fruits of *Delliniaindica*. L. *Sci. Asia., 33,* 57–60.

Kök, M., Hill, S., & Mitchell, J., (1999). A comparison of the rheological behavior of crude and refined locust bean gum preparations during thermal processing. *Carbohydr. Polym., 38,* 261–265.

Kumar, R., Patil, M., Patil, S., & Paschapur, M., (2009). Polysaccharides based colon-specific drug delivery: A review. *Int. J. Pharm. Tech. Res., 1,* 334–346.

Laurienzo, P., (2010). Marine polysaccharides in pharmaceutical applications: An overview. *Mar. Drugs., 8,* 2435–2465.

Lilia, R. E., & Morilla, M. J., (2011). Topical and mucosal liposomes for vaccine delivery. Wiley Interdisc. *Rev. Nanomed. Nanobiotechnol., 3,* 356–375.

Lundin, L., & Hermansson, A. M., (1995). Influence of locust bean gum on the rheological behavior and microstructure of K-*k*-carrageenan. *Carbohydr. Polym., 28,* 91–99.

Malafaya, P., Silva, G., & Reis, R., (2007). Natural–origin polymers as carriers and scaffolds for biomolecules and cell delivery in tissue engineering application. *Adv. Drug Deliv. Rev., 59,* 207–233.

Malakar, J., & Nayak, A. K., (2013). Floating bioadhesive matrix tablets of ondansetron HCl: Optimization of hydrophilic polymer-blends. *Asian J. Pharm., 7,* 174–183.

Malakar, J., Das, K., & Nayak, A. K., (2014). *In situ cross-linked matrix tablets for sustained salbutamol sulfate release – formulation development* by *statistical optimization. Polym. Med., 44,* 221–230.

Malakar, J., Nayak, A. K., & Goswami, S., (2012). Use of response surface methodology in the formulation and optimization of bisoprolol fumarate matrix tablets for sustained drug release. *ISRN Pharm., Article ID 730628.*

Malakar, J., Nayak, A. K., Jana, P., & Pal, D., (2013). Potato starch-blended alginate beads for prolonged release of tolbutamide: Development by statistical optimization and in vitro characterization. *Asian J. Pharm., 7*, 43–51.

Malik, K., Arora, G., & Singh, I., (2011a). Locust bean gum as superdisintegrant – Formulation and evaluation of nimesulide orodispersible tablets. *Polym. Med., 4*, 17–28.

Malik, K., Arora, G., & Singh, I., (2011b). Taste masked microspheres of ofloxacin: Formulation and evaluation of orodispersible tablets. *Sci. Pharm., 79*, 653–672.

Marianecci, C., Carafa, M., Di, M. L., Rinaldi, F., Di, M. C., & Alhaique, F., (2011). New vesicle-loaded hydrogel system suitable for topical applications: Preparation and characterization. *J. Pharm. Sci., 14*, 336–346.

Mathur, V., & Mathur, N., (2005). Fenugreek and other less known legume galactomannan polysaccharides: Scope for developments. *J. Sci. Ind. Res., 64*, 475–481.

Moin, A., & Shivakumar, H., (2010). Formulation and *in vitro* evaluation of sustained-release tablet of diltiazem: Influence of hydrophilic gums blends. *J. Pharm. Res., 3*, 600–604.

Mujoriya, R., Dhamande, K., Wankhede, U., & Angure, S., (2011). A review on study of buccal drug delivery system. *Innov. Syst. Des. Eng., 2*, 1–13.

Nakajima, N., & Matsuura, Y., (1997). Purification and characterization of konjac glucomannan degrading enzyme from anaerobic human intestinal bacterium, *Clostridium butyricum-Clostridium beijerinckii* group. *Biosci. Biotechnol. Biochem., 61*, 1739–1742.

Nayak A. K., (2016). Tamarind seed polysaccharide-based multiple-unit systems for sustained drug release, In: Kalia, S., & Averous, L., (eds.), *Biodegradable and Bio-Based Polymers: Environmental and Biomedical Applications* (pp. 471–494). Wiley-Scrivener, USA.

Nayak, A. K., & Malakar, J., (2010). Formulation and *in vitro* evaluation of gastroretentive hydrodynamically balanced system for ciprofloxacin HCl. *J. Pharm. Educ. Res., 1*, 65–68.

Nayak, A. K., & Malakar, J., (2011). Formulation and in vitro evaluation of hydrodynamically balanced system for theophylline delivery. *J. Basic Clin. Pharm., 2*, 133–137.

Nayak, A. K., & Manna, K., (2011). Current developments in orally disintegrating tablet technology. *J. Pharm. Educ. Res., 2*, 24–38.

Nayak, A. K., & Pal, D., (2012). Natural polysaccharides for drug delivery in tissue engineering. *Everyman's Sci., XLVI*, 347–352.

Nayak, A. K., & Pal, D., (2013a). Blends of jackfruit seed starch-pectin in the development of mucoadhesive beads containing metformin HCl. *Int. J. Biol. Macromol., 62*, 137–145.

Nayak, A. K., & Pal, D., (2013b). Ionotropically-gelled mucoadhesive beads for oral metformin HCl delivery: Formulation, optimization and antidiabetic evaluation. *J. Sci. Ind. Res., 72*, 15–22.

Nayak, A. K., & Pal, D., (2013c). Formulation optimization of jackfruit seed starch-alginate mucoadhesive beads of metformin HCl. *Int. J. Biol. Macromol., 59*, 264–272.

Nayak, A. K., & Pal, D., (2014). Trigonellafoenum-graecum L. seed mucilage-gellan muco-adhesive beads for controlled release of metformin HCl. *Carbohydr. Polym., 107*, 31–40.

Nayak, A. K., & Pal, D., (2015). Chitosan-based interpenetrating polymeric network systems for sustained drug release. In: Tiwari, A., Patra, H. K., & Choi, J. W., (eds.), *Advanced Theranostics Materials* (pp. 183–208). Wiley-Scrivener, USA.

Nayak, A. K., & Pal, D., (2016a). Plant-derived polymers: Ionically gelled sustained drug release systems, In: Mishra, M., (ed.), *Encyclopedia of Biomedical Polymers and Polymeric Biomaterials* (Vol. VIII, pp. 6002–6017). Taylor & Francis Group, New York, NY 10017, USA.

Nayak, A. K., & Pal, D., (2016b). Sterculia gum-based hydrogels for drug delivery applications. In: Kalia, S., (ed.), *Polymeric Hydrogels as Smart Biomaterials* (pp. 105–151). Springer series on polymer and composite materials, Springer International Publishing, Switzerland.

Nayak, A. K., & Pal, D., (2017a). Natural starches-blended ionotropically-gelled microparticles/beads for sustained drug release. In: Thakur, V. K., Thakur, M. K., & Kessler, M. R., (eds.), *Handbook of Composites from Renewable Materials* (Vol. 8, pp. 527–560). Nanocomposites: Advanced applications, Wiley-Scrivener, USA.

Nayak, A. K., & Pal, D., (2017b). Tamarind seed polysaccharide: An emerging excipient for pharmaceutical use. Indian. *J. Pharm. Educ. Res., 51*, S136–S146.

Nayak, A. K., & Pal, D., (2018). Functionalization of tamarind gum for drug delivery. In: Thakur, V. K., & Thakur, M. K., (eds.), *Functional Biopolymers, Springer International Publishing* (pp. 35–56). Switzerland.

Nayak, A. K., Ara, T. J., Hasnain, M. S., & Hoda, N., (2018a). Okra gum-alginate composites for controlled releasing drug delivery, In: Inamuddin, A. M., & Asiri, A. M., (eds.), *Applications of Nanocomposite Materials in Drug Delivery* (pp. 761–785). A volume in Woodhead Publishing Series in Biomaterials, Elsevier Inc.

Nayak, A. K., Beg S., Hasnain, M. S., Malakar, J., & Pal, D., (2018d). Soluble starch-blended Ca^{2+}-Zn^{2+}-alginate composites-based microparticles of aceclofenac: Formulation development and *in vitro* characterization. *Future J. Pharm. Sci., 4*, 63–70.

Nayak, A. K., Bera H., Hasnain, M. S., & Pal, D., (2018b). Graft-copolymerization of plant polysaccharides. In: Thakur, V. K., (ed.), *Biopolymer Grafting, Synthesis and Properties* (pp. 1–62). Elsevier Inc.

Nayak, A. K., Das, B., & Maji, R., (2012a). Calcium alginate/gum Arabic beads containing glibenclamide: Development and *in vitro* characterization. *Int. J. Biol. Macromol., 51*, 1070–1078.

Nayak, A. K., Hasnain, M. S., & Pal, D., (2018c). Gelled microparticles/beads of sterculia gum and tamarind gum for sustained drug release. In: Thakur, V. K., & Thakur, M. K., (eds.), *Handbook of Springer on Polymeric Gel* (pp. 361–414). Springer International Publishing, Switzerland.

Nayak, A. K., Hasnain, M. S., Beg, S., & Alam, M. I., (2010a). Mucoadhesive beads of gliclazide: Design, development and evaluation. *Sci. Asia., 36*, 319–325.

Nayak, A. K., Kalia, S., & Hasnain, M. S., (2013a). Optimization of aceclofenac-loaded pectinate-poly (vinyl pyrrolidone) beads by response surface methodology. *Int. J. Biol. Macromol., 62*, 194–202.

Nayak, A. K., Khatua, S., Hasnain, M. S., & Sen, K. K., (2011). Development of alginate-PVP K 30 microbeads for controlled diclofenac sodium delivery using central composite design. *DARU J. Pharm. Sci., 19*, 356–366.

Nayak, A. K., Maji, R., & Das, B., (2010b). Gastroretentive drug delivery systems: A review. *Asian J. Pharm. Clin. Res., 3*, 2–10.

Nayak, A. K., Pal, D., & Das, S., (2013b). Calcium pectinate-fenugreek seed mucilage mucoadhesive beads for controlled delivery of metformin HCl. *Carbohydr. Polym., 96,* 349–357.

Nayak, A. K., Pal, D., & Hasnain, M. S., (2013c). Development and optimization of jack-fruit seed starch-alginate beads containing pioglitazone. *Curr. Drug Deliv., 10,* 608–619.

Nayak, A. K., Pal, D., & Malakar, J., (2013e). Development, optimization and evaluation of emulsion-gelled floating beads using natural polysaccharide-blend for controlled drug release. *Polym. Eng. Sci., 53,* 338–350.

Nayak, A. K., Pal, D., & Santra, K., (2013d). Plantago ovata F. Mucilage-alginate mucoadhesive beads for controlled release of glibenclamide: Development, optimization, and *in vitro-in vivo* evaluation. *J. Pharm., Article ID 151035.*

Nayak, A. K., Pal, D., & Santra, K., (2014a). *Artocarpusheterophyllus L.* seed starch-blended gellan gum mucoadhesive beads of metformin HCl. *Int. J. Biol. Macromol., 65,* 329–339.

Nayak, A. K., Pal, D., & Santra, K., (2014b). Development of calcium pectinate-tamarind seed polysaccharide mucoadhesive beads containing metformin HCl. *Carbohydr. Polym., 101,* 220–230.

Nayak, A. K., Pal, D., & Santra, K., (2014c). Development of pectinate-ispagula mucilage mucoadhesive beads of metformin HCl by central composite design. *Int. J. Biol. Macromol., 66,* 203–221.

Nayak, A. K., Pal, D., & Santra, K., (2014d). Ispaghula mucilage-gellan mucoadhesive beads of metformin HCl: Development by response surface methodology. *Carbohydr. Polym., 107,* 41–50.

Nayak, A. K., Pal, D., & Santra, K., (2014e). Tamarind seed polysaccharide-gellan mucoadhesive beads for controlled release of metformin HCl. *Carbohydr. Polym., 103,* 154–163.

Nayak, A. K., Pal, D., & Santra, K., (2015). Screening of polysaccharides from tamarind, fenugreek and jackfruit seeds as pharmaceutical excipients. *Int. J. Biol. Macromol., 79,* 756–760.

Nayak, A. K., Pal, D., & Santra, K., (2016). Swelling and drug release behavior of metformin HCl-loaded tamarind seed polysaccharide-alginate beads. *Int. J. Biol. Macromol., 82,* 1023–1027.

Nayak, A. K., Pal, D., Pany, D. R., & Mohanty, B., (2010c). Evaluation of *Spinacia oleracea L.* leaves mucilage as innovative suspending agent. *J. Adv. Pharm. Technol. Res., 1,* 338–341.

Nayak, A. K., Pal, D., Pradhan J., & Hasnain, M. S., (2013f). Fenugreek seed mucilage-alginate mucoadhesive beads of metformin HCl: Design, optimization and evaluation. *Int. J. Biol. Macromol., 54,* 144–154.

Nayak, A. K., Pal, D., Pradhan, J., & Ghorai, T., (2012b). The potential of *Trigonellafoenum-graecum L.* seed mucilage as suspending agent. *Indian J. Pharm. Educ. Res., 46,* 312–317.

Pal, D., & Nayak, A. K., (2015a). Alginates, blends and microspheres: Controlled drug delivery, In: Mishra, M., (ed.), *Encyclopedia of Biomedical Polymers and Polymeric Biomaterials* (Vol. I, pp. 89–98). Taylor & Francis Group, New York, NY 10017, USA.

Pal, D., & Nayak, A. K., (2015b). Interpenetrating polymer networks (IPNs): Natural polymeric blends for drug delivery, In: Mishra, M., (ed.), *Encyclopedia of Biomedical Polymers and Polymeric Biomaterials* (Vol. VI, pp. 4120–4130). Taylor & Francis Group, New York, NY 10017, USA.

Pal, D., & Nayak, A. K., (2017). Plant polysaccharides-blended ionotropically-gelled alginate multiple-unit systems for sustained drug release, In: Thakur, V. K., Thakur, M. K., & Kessler, M. R., (eds.), *Handbook of Composites from Renewable Materials* (Vol. 6, pp. 399–400). Polymeric Composites, Wiley-Scrivener, USA.

Pal, D., Nayak, A. K., & Kalia, S., (2010). Studies on Basella alba L. leaves mucilage: Evaluation of suspending properties. *Int. J. Drug Discov. Tech., 1*, 15–20.

Pal, D., Nayak, A. K., & Saha S., (2018). Interpenetrating polymer network hydrogels of chitosan: Applications in controlling drug release, In: *Mondal, I. H., (ed.), Cellulose-Based Superabsorbent Hydrogels, Polymers and Polymeric Composites: A Reference Series* (pp. 1–41). Springer, Cham.

Patel, M., Tekade, A., Gattani, S., & Surana, S., (2008). Solubility enhancement of lovastatin by modified locust bean gum using solid dispersion techniques. *AAPS Pharm. Sci. Tech., 9*, 1262–1269.

Pollard, M., Kelly, R., Wahl, C., Fischer, K. P., Windhab, E., & Eder, B., (2007). Investigation of equilibrium solubility of a carob galactomannan. *Food Hydrocol., 21*, 683–692.

Prajapati, V. D., Jani, G. K., Moradiya, N. G., Randeria, N. P., & Nagar, B. J., (2013). Locust bean gum: A versatile biopolymer. *Carbohydr. Polym., 94*, 814–821.

Rath, A. S. N., Nayak, B. S., Nayak, A. K., & Mohanty, B., (2010). Formulation and evaluation of buccal patches for delivery of atenolol. *AAPS Pharm. Sci. Tech., 11*, 1034–1044.

Ray, S., Sinha, P., Laha, B., Maiti, S., Bhattacharyya, U. K., & Nayak, A. K., (2018). Polysorbate 80 coated crosslinked chitosan nanoparticles of ropinirole hydrochloride for brain targeting. *J. Drug Deliv. Sci. Tech., 48*, 21–29.

Rowe, R., Sheskey, P., & Owen, S., (2006). *Handbook of Pharmaceutical Excipients* (5th edn.). London: Pharmaceutical Press.

Ruiz-Roso, B., Quintela, J., De la Fuente E., Haya, J., & Pérez-Olleros, L., (2010). Insoluble carob fiber rich in polyphenols lowers total and LDL cholesterol in hypercholesterolemic sujects. *Plant Foods Human Nut., 65*, 50–56.

Sandolo, C., Bulone, D., Mangione, M. R., Margheritelli, S., Di Meo, C., Alhaique, F., et al., (2010). Synergistic interaction of locust bean gum and xanthan investigated by rheology and light scattering. *Carbhydr. Polym., 82*, 733–741.

Sandolo, C., Coviello, T., Matricardi, P., & Alhaique, F., (2007). Characterization of polysaccharide hydrogels for modified drug delivery. *Eur. Biophys. J., 36*, 693–700.

Şenel, S., (2010). Potential applications of chitosan in oral mucosal delivery. *J. Drug Deliv. Sci. Technol., 20*, 23–32.

Silveira, J. L. M., & Bresolin, T. M. B., (2011). Pharmaceutical use of galactomannans. *Quim. Nova, 34*, 292–299,

Sinha, M. S., Mohanta, S., & Nayak, A. K., (2011). Preliminary investigation on angiogenic potential of Ziziphusoenoplia M. root ethanolic extract by chorioallantoic membrane model. *Sci. Asia., 37*, 72–74.

Sinha, P., Ubaidulla, U., & Nayak, A. K., (2015). Okra (*Hibiscus esculentus*) gum-alginate blend mucoadhesive beads for controlled glibenclamide release. *Int. J. Biol. Macromol., 72* (2015b), 1069–1075.

Sinha, P., Ubaidulla, U., Hasnain, M. S., Nayak, A. K., & Rama, B., (2015). Alginate-okra gum blend beads of diclofenac sodium from aqueous template using $ZnSO_4$ as a cross-linker. *Int. J. Biol. Macromol., 79* (2015a), 555–563.

Sinha, V., & Kumria, R., (2001). Polysaccharides in colon-specific drug delivery. *Int. J. Pharm., 224*, 19–38.

Sinico, C., & Fadda, A. M., (2009). Vesicular carriers for dermal drug delivery. *Expert Opin. Drug Deliv., 6*, 813–825.

Smith, B., Bean, S., Schober, T., Tilley, M., Herald, T., & Aramouni, F., (2010). Composition and molecular weight distribution of carob germ protein fractions. *J. Agri. Food Chem., 58*, 7794–7800.

Soumya, R., Ghosh, S., & Abraham, E. T., (2010). Preparation and characterization of guar gum nanoparticles. *Int. J. Biol. Macromol., 46*, 267–269.

Srivastava, M., & Kapoor, V., (2005). Seed galactomannans: An overview. *Chem. Biodiv., 2*, 295–317.

Sujja-areevath, J., Munday, D., Cox, P., & Khan, K., (1998). Relationship between swelling, erosion and drug release in hydrophilic natural gum mini-matrix formulations. *Eur. J. Pharm. Sci., 6*, 207–217.

Suzuki, S., & Lim, J. K., (1994). Microencapsulation with carrageenan-locust bean gum mixture in a multiphase emulsification technique for sustained drug release. *J. Microencapsul., 11*, 197–203.

Urdiain, M., Doménech-Sánchez, A., Albertí, S., Benedí, V., & Rosselló, J., (2004). Identification of two additives, locust bean gum (E-410) and guar gum (E- 412), in food products by DNA-based methods. *Food Add. Contamin., 21*, 619–625.

Venkataraju, M. P., Gowda, D. V., Rajesh, K. S., & Shiva, K. H., (2007). Xanthan and locust bean gum matrix tablets for oral controlled delivery of propranolol HCl. *Asian J. Pharm. Sci., 2*, 239–248.

Vijayaraghavan, C., Vasanthakumar, S., & Ramakrishnan, A., (2008). *In vitro* and *in vivo* evaluation of locust bean gum and chitosan combination as a carrier for buccal drug delivery. *Pharmazie., 63*, 342–347.

Watanabe, K., Yakou, S., Takayama, K., Machida, Y., & Nagai, T., (1992). Factors affecting prednisolone release from hydrogels prepared with water-soluble dietary fibers, xanthan and locust bean gums. *Chem. Pharm. Bull., 40*, 459–462.

Yamagar, M., Kadam, V., & Hirlekar, R., (2010). Design and evaluation of buccoadhesive drug delivery system of metoprolol tartrate. *Int. J. Pharm. Tech. Res., 2*, 453–462.

Zavoral, J., Hannan, P., Fields, D., Hanson, M., Frantz, I., & Kuba, K., (1983). The hypolipidemic effect of locust bean gum food products in familial hypercholesterolemic adults and children. *Am. J. Clin. Nutr., 38*, 285–294.

Zunft, H. J., Lüder, W., Harde, A., Haber, B., Graubaum, H. J., Koebnick, C., et al., (2003). Carob pulp preparation rich in insoluble fiber lowers total and LDL cholesterol in hypercholesterolemic patients. *Eur. J. Nutr., 42*, 235–242.

CHAPTER 7

Pharmaceutical Applications of Sterculia Gum

MD NURUS SAKIB, MD MINHAJUL ISLAM, MD SHAHRUZZAMAN,
ABUL K. MALLIK, PAPIA HAQUE, and
MOHAMMED MIZANUR RAHMAN

*Department of Applied Chemistry and Chemical Engineering, Faculty
of Engineering and Technology, University of Dhaka, Dhaka-1000,
Bangladesh*

ABSTRACT

Sterculia gum, also known as Karaya gum, is an exudate by *Sterculia* trees. Sterculia gum is a natural acid polysaccharide and yields galactose, rhamnose, and galacturonic acid after hydrolysis. Sterculia gum use has long been in the history as a medicinal remedy. Having potential for the improvement of drugs owing to its chemical, biological, and physical properties, the Karaya gum is a valuable product and is in use mostly in the fields of pharmaceuticals, food additives, denture adhesive and adhesive for oral treatment. There has been an increasing interest for the modification of Sterculia gum for enhancing the solubility of drugs and drug carrier characteristics. Like Xanthan gum, Sterculia gum is also used as a controlling agent for drug release. Other applications for Sterculia gum are the treatment of Verruca vulguris, enhancement of sealant, oral medicament auxiliary, etc. In this chapter, pharmaceutical applications of Sterculia gum will be discussed in detail.

7.1 INTRODUCTION

Gums, pathological products of plants, are polysaccharide polymers, which are formed due to unfavorable conditions or breakage of cell walls of a plant. Monosaccharide units of natural gums are linked with glucosidic bonds. Natural exudate gums have been a part of the history of mankind from about 5000 years ago and were used as a thickener at that time (Phillips and Williams, 2001). Gum Arabic (GA), Gum Tragacanth (GT) and Sterculia or Karaya gum are one of the most commonly used gums nowadays.

Natural gums are biodegradable. They do not cause any irritation or any adverse effects, i.e., non-toxic to human or nature owing to their carbohydrates bases. On the other hand, the use of synthetic gums may cause acute effects. For example, handling substances such as methyl methacrylate may result in chronic adverse effects (Rowe, 2009). Some of the synthetic gums may cause local or systemic reactions as well. For example, a mild inflammatory response has been seen in rats in implant studies using poly-(propylene fumarate) (Jani et al., 2009). But natural gums are biocompatible due to their natural origin and safe to use.

Sterculia gum or karaya gum is the exudation from the *Sterculia* sp. or from *Cochlospermum* sp. It is a complex calcium or magnesium salts of polysaccharide. Commercial karaya gum contains mostly uronic acid residues while a little fraction of acetyl groups. Presence of such acetyl groups makes the gum insoluble to water. However, after de-acetylation, it becomes water soluble. Sterculia gum obtained from different sources and species tend to maintain their chemical compositions (Whistler, 1993). Karaya gum is defined by JECFA as a dried exudation which sources from the stems and branches of Sterculia urens Roxburg and other species of Sterculia of family Sterculiaceae or from Cochlospermum gossypium A.P. De Candolle or species of the same genus of family Bixaceae (Verbeken et al., 2003).

Karaya gum was introduced in the market in the early 1900s to add as an adulterant for GT for the similarity. However, karaya gum is now a commercially important item for the use in the fields of pharmaceutical, adhesive, gelling agent, textile, etc. Largest producer and exporter of karaya gum is India, and they exported around 4000–6000 tons in an average from 1960–1980. Presently, the export of such gum is increasing from the African market. In Africa, Senegal is the largest exporter; exporting over 1000 tons annually. Among the importers, France, and the United Kingdom are at the top with an annual import of around 400 tons and 210 tons, respectively (Singh, 2010; Verbeken et al., 2003). The price for karaya gum is around US $2250/tons to US $6000/tons (Verbeken et al., 2003).

Natural gums like Sterculia, i.e., karaya gum possess high cohesion and adhesion characteristics owing to their branched structure. These characteristics are sought highly in pharmaceutical industries, and hence natural gums are now in the spotlight on how to use them more effectively as a suspending agent, dental adhesive, tablet binder, laxative, etc. This chapter will discuss the pharmaceutical applications of sterculia gum and modified sterculia gums.

7.2 SOURCES AND PREPARATION OF STERCULIA GUM

Natural gums are complex carbohydrate polysaccharides extracted from different sources such as endosperm of plant seeds (guar gum), plant exudates (tragacanth (Trag)), tree or shrub exudates (GA, Sterculia gum) seaweed extracts (agar), bacteria (xanthan gum), and animal sources (chitin), etc. Sterculia gum, an important tree gum, originates from *Sterculia urens* tree, a large, bushy deciduous tree that can grow up to 15 m high, which belongs to *Sterculiaceae* family. This is cultivated in India and usually grows in dry and rocky forests regions (Galla and Dubasi, 2010). *Sterculia urens* is indigenous to central and northern India, and the majority of this gum is produced in the state of Andhra Pradesh. Other noted sources are *S. setigera* in Senegal and Mali and *S. villosa* in Sudan, India, and Pakistan.

Sterculia gum or Karaya gum is a complex, partially acetylated polysaccharide obtained as a calcium and magnesium salt. The karaya gum collection starts after wounding the Sterculia tree. The exudation is higher in the first few days. The exude solidifies into the form of drops when dried. Dried gums are then collected. Generally, a mature tree gives around 1–5 kg of gum per season. The gum collection is best from April to June. It is prepared by first removing impurities such as bark, stones, fibers, and sand by manually or mechanically. Then it is then milled, blended. Finally, the gum is classified on the basis of mesh size, viscosity, and purity. Commercial karaya gum can be of five grades depending on the quality. The gum is available as granules or in powder form. The color of the powder may be light to pinkish gray and has a slight acetic taste and odor. The higher grade is of a lighter color. The granule size ranges from 4–8 mesh and 8–14 mesh and powder size is 160 mesh with a viscosity ranging from 500–1200 cps (López-Franco et al., 2009; Verbeken et al., 2003).

7.3   PHYSICAL AND CHEMICAL PROPERTIES OF STERCULIA GUM

Sterculia gum is a branched acidic polysaccharide. The molecular mass of karaya gum is around 16×10^6 Da. The major constituent of Sterculia gum is d-galacturonic acid, d-galactose, L-rhamnose, and d-glucuronic acid. The structure of Sterculia gum is composed of rhamnogalacturonan which has α-(1-4)-linked D-galacturonic acid and α-(1-2)-linked-L-rhamnosyl residues in the main chain, whereas the side chain contains (1-3)-linked β-D-glucuronic acid, or (1-2)-linked β-D-galactose on the galacturonic acid unit. Here one half of the rhamnose is substituted with (1-4)-linked β-D-galactose. (Mittal et al., 2015) When partially hydrolyzed, the karaya gum gives monosaccharides and acidic oligosaccharides (Figures 7.1 and 7.2). The oligosaccharides contain mostly 2-D-galacturonosyl-L-rhamnose and are 4-D-galacturonosyl-D-galactose (Hirst and Dunstan, 1953).

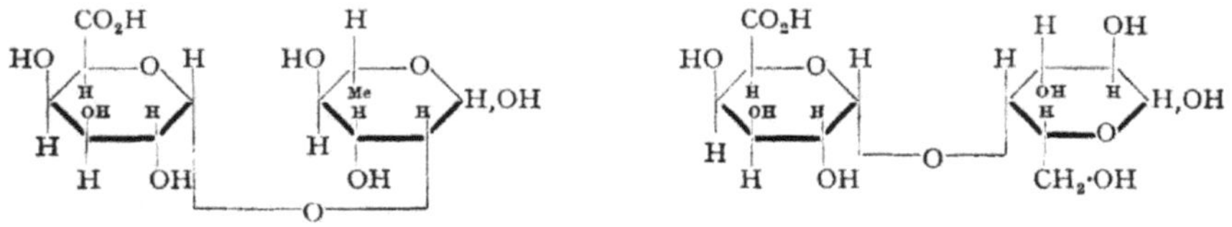

FIGURE 7.1   Products obtained from partial hydrolysis of Karaya gum: (I) is 2-D-galacturonosyl-L-rhamnose, and (II) is 4-D-galacturonosyl-D-galactose. Reprinted with permission from Hirst and Dunstan (1953). © Royal Society of Chemistry.

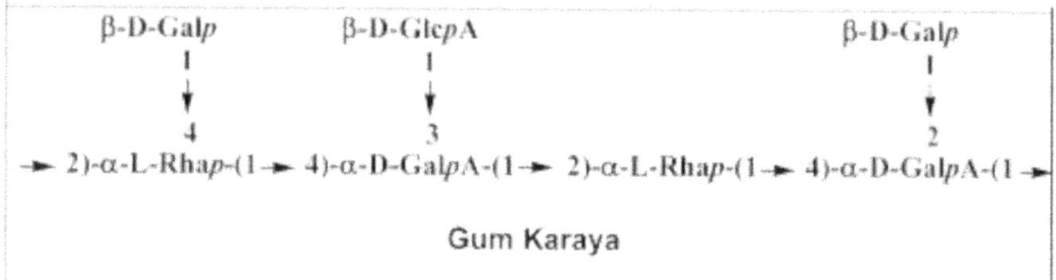

FIGURE 7.2   Structure of Karaya gum. Reprinted with permission from Rana et al., (2011). © Elsevier.

Sterculia Urens' seed contains 56% kernel, which has 35% protein, 26% oil, and 28% carbohydrates (Galla and Dubasi, 2010). The chemical structure of Sterculia gum also contains aspartic acid (64.2±2.44 µg/g), proline (30.5±1.86 µg/g), glutamic acid (34.2± 1.44 µg/g), threonine (25.2±1.06 µg/g) and glycine (4.8±0.45 µg/g) and leucine (3.9±0.28 µg/g). Different types of saturated and unsaturated fatty acids were also found in the structure of Sterculia gum. Fatty acids such as stearic acid (25.5±1.64 µg/g), palmitic acid (18.5±0.95 µg/g), palmitoleic acid (13.2±0.95 µg/g), lauric acid (12.8±0.62 µg/g) and oleic acid (4.2±0.21 µg/g).

Sterculia gum has the lowest solubility in water compared to other commercial gums. This gum produces true solutions only at very low concentrations (<0.02% in cold water, 0.06% in hot) (Le Cerf et al., 1990). But it is possible to prepare highly viscous colloidal dispersions at concentrations up to 5%, but that depends on the quality of the gum. The insolubility of Sterculia gum in water can be attributed to the presence of the acetyl group on the structure. This is also the reason behind Sterculia gum's inability to produce a clear solution, rather this gum produces viscous colloid in the water at low concentration. While in 60% alcohol, Sterculia gum produces viscous solution at low concentration but, like in water, is insoluble in alcohol at high concentration. Water insolubility of Sterculia gum can be tackled by de-acetylation by using alkali (Le Cerf et al., 1990). A smooth and homogeneous gum can be prepared with fine mesh gum, but coarser granules produce grainy dispersion. In general, Sterculia gum from Indian sources (mainly from *S. urens*) has a higher acid value and a distinct acetic odor than that of African origin (mainly *S. setigera*). This results in African Sterculia gum having a better solubility than Indian Sterculia gum (López-Franco et al., 2009).

The viscosity of Sterculia gum dispersion ranges from about 120–400 centipoise (cPs) for 0.5% dispersions to about 10,000 cPs for 3% dispersions. Viscosity is near to infinity at low shear rate values; moreover, these gum exhibits yield stresses of 60 and 100 micro N/cm^2 respectively at concentrations of 2.0 and 3.0% in water (Mills and Kokini 1984). As a result, concentrated gum solutions are able to suspend particles and give soft, spreadable gels with a jam-like consistency (Imeson, 2012).

The smoothness of the gum solution is governed by the particle size. This can be altered by continuous stirring, which leads to a smooth texture and reduced viscosity (Nussinovitch, 1997). Sterculia gum loses viscosity on aging in the dry state and builds an acetic odor. The loss of viscosity can be attributed to the loss of acetic acid. In fact, finely powdered gum shows higher loss of viscosity than granules or whole exudate. This drop in viscosity is most evident in the first few weeks after the gum has been ground. The stability of the gum can be affected by the high temperature of high stability. So, it is important to store the gum at a temperature that does not exceed 25°C (Mortensen et al., 2016). The viscosity of Sterculia gum can be affected by temperature and pH of the solution, Climate, and time of harvest and presence of electrolytes. Sterculia gum is more viscous when hydrated in the cold rather than in hot water. Viscosity can be reduced if the gum is subjected to boiling temperature for more than two minutes. The viscosity of Sterculia gum decreases with the addition of an acid or alkali. The viscosity of Sterculia gum may decrease with the addition of electrolytes as well. The dispersion is not affected by weak electrolytes, but when certain strong electrolytes are added, even in small amount, loss of viscosity is observed (Whistler, 1993).

Sterculia gum shows pH in the range of 4.5–4.7 for Indian origin and 4.7–5.2 for African origin (1% solution). At higher pH, mainly above pH 8, Sterculia gum is converted into a ropy and stringy mucilage due to loss of acetyl groups from molecules as a result of rapid saponification. As a result of the presence of high uronic acid content, Sterculia gum dispersions bear acid conditions really well and resist hydrolysis in 10% hydrochloric acid solution at room temperature for at least 8 hours (Whistler, 1993).

7.4 ADVANTAGE OF USING NATURAL GUMS

7.4.1 BIODEGRADABLE

Natural gums are sourced from nature, and hence, they are easily biode-gradable and produce zero waste.

7.4.2 RENEWABLE SOURCE

Natural gums are collected from the barks of trees. Hence, the economic cultivation of such trees is often possible. Since the sources are renewable, natural gums can have a sustainable production with often locally available sources.

7.4.3 BIOCOMPATIBLE

Natural gums do not show any adverse effects. Food and Drug Administration (FDA) has given GRAS (generally recognized as safe) status to Sterculia gum after toxicological, teratological, and mutagenic tests proved its (Mbuna and Mhinzi, 2003). It was reported by a number of studies that Sterculia gum is safe for food owing to the fact that it is neither digested nor degraded by enteric microflora or adsorbed to any significant extent in human beings (Eastwood et al., 1983).

7.5 PHARMACEUTICAL APPLICATIONS OF STERCULIA GUM

7.5.1 PURGATIVE USE

Sterculia gum is a hydrophilic complex polysaccharide which shows rapid swelling from 60 up to 100 times of their original volume. Since sterculia is non-toxic to humans and does not get digested in the intestines, the potential for it to become a bulk laxative is high. After psyllium seed, Sterculia gum has the most importance as a bulk laxative. Due to the swelling, it gives a mechanical distention for evacuation. Furthermore, the swelling is discontinuous, which is thought to be very effective as a laxative (Meier et al., 1990). A study compared Sterculia gum to Isapgol husk on the swelling index, granular friability, etc., characteristics, and found that sterculia gum had similar properties like the laxative formulations found in the market. Since it does not cause any intestinal problems, it was suggested that people suffering from habitual constipation might use sterculia as a bulk laxative (Harshal and Priscilla, 2011).

Sterculia gum was used to combat chronic constipation in elderly patients (Meier et al., 1990). Sterculia gum has the capacity to absorb a large quantity of water and does not disintegrate considerably in the alimentary tract. As a result, daily administration of laxative Sterculia gum was more effective in managing chronic constipation.

7.5.2 BIO-ADHESION MATERIAL

Sterculia gum has also exhibited applicability as adhesive for dental fixtures and ostomy equipment. Accumulation of denture plaque can be prevented by coating with Sterculia gum. Moreover, other problems associated with dentures such as stomatitis, staining, and bad breath can be tackled by Sterculia gum coating (Bart et al., 1989; Wilson and Harvey, 1989).

Sterculia gum, i.e., Karaya gum as an adhesive on dental plates showed some advantages compared to other adhesives. When in contact with the mouth, karaya gum swells up and give a comfortable and tighter adjustment to the plate. Owing to its own anti-bacterial property, it also resists bacterial degradation (Steinhardt and Goldwater, 1962).

7.5.3 DRUG DELIVERY

Natural gums have been traditionally in use in the fields of food technology and pharmaceuticals (Jani et al., 2009). Earlier, excipients were used to give proper weight, consistency, and volume to the drug formulation (Tekade and Chaudhari, 2013). Presently, they are being modified and blended with other compounds due to their bioavailability, stability, and low fabrication cost. Thus they are being exploited by the researchers to compete with the synthetic excipients (Tekade and Chaudhari, 2013).

For drug administration, the oral route is the most preferred route since it is convenient and more economical. Presently, fast disintegrating tablets are formulated using gum karaya, modified starch, and agar, all of which are of natural origin. Karaya gum has found its use in release controlling agents, mucoadhesive tablets, super disintegrants, etc. Here sterculia gums, i.e., Karaya gums use as a drug carrier, controlling agent, disintegrant are discussed along with experimental findings.

Compressed matrices using karaya gum was produced to study its efficacy as a release controlling agent. Caffeine and diclofenac sodium (DS) was used as the model drugs, and the result indicated that karaya gum showed zero order drug release with erosion acting as the main cause for the release (Munday and Cox, 2000).

Gum Karaya or Sterculia gum has been used as a buccoadhesive for the controlled release of amoxicillin trihydrate, a semi-synthetic antibiotic. The researchers used different combinations with different polymers to find the best suitable polymer on the basis of several physicochemical parameters. A tablet using gum karaya, the excipient, and the drug were prepared by means of wet granulation method. Different compositions of karaya gum and the drug were tested that gave different results. For 20% karaya gum mixture, the buccoadhesion strength was found to be the most. Generally, the adhesion between the layers was found to increase gradually to an optimum value with the increase of the degree of hydration. The drug release rate was found to vary as well. For 20% karaya gum concentration, the release rate T50% was the slowest, around 10 hours and 2 minutes, whereas for 10% the value is only 2 hours and 35 minutes. The release rate is dependent on the amount of gum used. In general, the higher amount of karaya gum provides maximum swelling, which eventually controls the drug release. However, it was found that drug was totally released within 24 hours and in vitro permeation was found to be almost 50–70%. The drug release kinetics for 10, 12 and 15% were found to follow zero order and independent of concentration and hence regarded as a potential release of drugs in the buccal region (Biswas et al., 2014).

Sterculia gum can be used to make hydrogels. One research shows that PVA and PVA-poly (AAm) hydrogel cross-linked with sterculia showed better compatibility and potential for wound dressings. The hydrogel also showed non-Fickian and Case II diffusion mechanism of antibiotic drug release. It had good impermeability against microorganism while permeable to oxygen and water vapor. Among the two, the sterculia-PVA showed good blood compatibility than sterculia PVA-poly (AAm) while the latter adsorbed more fluid when applied as a wound dressing (Singh and Pal, 2012).

Hydrogels made from sterculia gum, have the potential to be a controllable and sustainable drug delivery medium. One study shows that sterculia-PVA hydrogel can be developed for drug delivery. By means of irradiation of different doses, sterculia-cl-poly (VA) hydrogels were prepared, and the drug release mechanism of the hydrogel showed non-Fickian diffusion mechanism (Singh et al., 2011). The use of sterculia gum in anti-ulcer drug delivery has been investigated and shows good potential. A hydrogel made from sterculia gum and polyacrylamide was used to measure the release behavior of Ranitidine hydrochloride. At pH 2.2, the drug release is the maximum compared to other different pH conditions. Half of the total drug was released by 47 minutes in distilled water, 59 min in pH 2.2 buffer and 88 minutes for pH 7.4 buffer. Drug-loaded hydrogel shows Fickian diffusion mechanism in low pH and Non-Fickian in pH 7.4 buffers (Singh and Sharma, 2008).

Drug release studies using natural gums as binders were compared by Prasanthi et al. Dissolution rates for Ziprasidone tablets of different natural gums such as acacia, Trag, guar gum, gum karaya, and gum olibanum were found out. Karaya gum showed a higher dissolution rate and drug release than any other gums. Furthermore, the drug was not affected by the gums so that natural gums can be used as a binder in this case (Prasanthi et al., 2011).

Another drug delivery experiment of *Sterculia foetida* gum was performed with diltiazem hydrochloride. The findings show that tablets prepared from the sterculia gum had higher sustainable release than tablets prepared with HPMC (hydroxypropyl methylcellulose) matrix tablets since they are widely used as a hydrophilic swellable matrix forming a polymer. Also, the drug release from the gum showed linear with time, whereas, for HPMC, the release rate profile followed Higuchi equation. Thus, sterculia gum could be a better alternative than HPMC for the release of diltiazem hydrochloride due to its linear release rate (Chivate et al., 2008). A combination of HPMC and karaya gum showed better-floating capability and shorter floating lag time during an 8h drug release period in another study (Afrasim and Shivakumar, 2010).

Sterculia gum for colon targeted drug delivery faces some difficulties due to its insolubility in water, quick hydration, and swelling into a homogenous mass hydrogel (Nath and Nath, 2012). Researchers have been done to check the suitability of sterculia gum as a colonic drug delivery carrier. The susceptibility of the sterculia gum in rat caecal microflora suggested that the sterculia gum gives premature drug release if not coated with enteric coating and may not be able to reach colonic area (Nath and Nath, 2013). However, researches of the coated drug prepared with sterculia gum being a hydrophilic matrix showed potential for colonic drug delivery. A microflora triggered colon targeted drug delivery system (MCDDS) was concocted using azathioprine as the core, sterculia gum as the binder and the provider of the hydrostatic pressure inside the tablet and finally, coating for providing resistance. The coating layer is comprised of chitosan (CS) and Eudragit RLPO polymer for intestinal resistance and controlling drug release. The release study of the drug to colon showed that the system was able to resist the gastric and intestinal condition but was vulnerable to bacterial attack (Nath and Nath, 2012).

Park and Mundays' research on preparing biocompatible tablets using xanthan gum, karaya gum, guar gum presented some important findings on the characteristics of karaya gum in comparison to others. Findings show that karaya gum showed better adhesion to mucosal membrane than guar gum. Karaya gum also showed better hydration and swelling than guar gum. In addition, researchers had also shown that with the increase in karaya gum's weight percentage in tablets, the drug release rate of Nicotine hydrogen tartrate decreases (Park and Munday, 2004).

For studying the mucoadhesive properties of sterculia or karaya gum and alginate, coated microcapsules were prepared by Rama Krishna et al. They prepared the Glipizide microcapsule by means of both ionic gelation process and emulsification ionotropic gelation process. The latter process found out to be more favorable due to the slow and total release of glipizide over time. Significant hypoglycemic effect and mucoadhesive properties make these capsules as a potential candidate for drug delivery (Krishna et al., 2009).

For poorly water-soluble drugs, Sterculia gum that is, gum karaya can enhance the dissolution rate. A modified gum karaya as the carrier was studied to check the dissolution rate of the drug nimodipine. The karaya gum was modified by heating to 120°C for more than 2 hours. The drug was mixed with karaya gum in the co-grinding fashion since it is comparatively easier and convenient according to the researchers. The dissolution rate of the drug increased with the increase in the modified karaya gum concentration. The rate is higher than the non-modified karaya gum as the carrier since the non-modified karaya gum showed high viscosity in the microenvironment of the drug, which eventually reduced the dissolution rate of the drug. Karya gum in modified form showed better dissolution rate while the crystallinity of the drug remained the same in both the carriers (Babu et al., 2002). Another example of karaya gum as the dissolution enhancer is the glimepiride dissolution enhancement using the modified karaya gum as the carrier. In this case, the drug was dispersed by means of solvent evaporation method (Nagpal et al., 2012). Another research showed a comparison between karaya gum and modified karaya gum in terms of dissolution time and dissolution efficiency of matrix tablets. Karaya gum was cross-linked by tri-sodium tri-metaphosphate, and the result shows that modified karaya gum showed higher mean dissolution time. About 68.2% drug was released from the modified karaya gum tablet whereas for unmodified karaya gum tablet, it was 99.9% by 10 hours (Reddy et al., 2012). A solid dispersant made of modified karaya gum and aceclofenac was applied as a matrix-forming agent for controlled oral release. Aceclofenac is an anti-inflammatory drug, which is often used to treat pain and arthritis (Arora et al., 2011).

Karaya gums can also act as super disintegrants. Immediate release tablets disintegrate very fast to deliver the drug rapidly. Disintegrants are important in tablet formulation. The choice of disintegrants is a critical factor. Disintegrant should be able to interact with water strongly. Use of karaya gum for the immediate release of drugs was found to be more effective for a number of cases. In one case, modified karaya gum and locust bean gum (LBG) were investigated as the superdisintegrants. Modified Karaya gum showed fast disintegration time and 100% release of the drug olanzapine. The dissolution rate was of the first order and linear. Among the two, the karaya gum showed better responses (Kalyani et al., 2014).

Another case shows the formulation of immediate release tablet of olanzapine drug that uses karaya gum and *Hibiscus rosa-sinensis* mucilage. The dissolution rate varied on the superdisintegrant's concentration. The tablet showed better disintegration than the available synthetic ones. Research with amlodipine besylate as a drug was conducted using the modified karaya gum and *Hibiscus rosa-sinensis* mucilage. The result was similar; gum karaya showed shorter disintegration time and 100% release (Sukhavasi and Kishore, 2012). Hence karaya gum can be a good superdisintegrant for drugs.

Karaya gum has found its use in ileostomy appliances as well. Marshal Sparberg et al., have shown that powdered gum karaya composed ring could be used both as a disposable or permanent appliance in an ileostomy. The disposable ring used, prevented post-operative skin problems as well as assisting in the healing of the surgery area. As a permanent appliance, karaya gum ring can be used as a substitute for adhesive. The authors point out that this is useful since some patients develop allergies to synthetic adhesives (Sparberg et al., 1966).

Karaya gum patch has been found effective for the treatment of verruca vulgaris. Karaya gum patch was used to administer salicylic acid by means of transdermal delivery technology. Around 69% cure rate was obtained using karaya gum patches (Bart et al., 1989).

7.6 OTHER USES

Sterculia gum has been used in pharmaceutical, paper, textile, bakery, dairy products, and fabrication of composites and hydrogel adsorbents. Microbial quality of this gum is satisfactory to use in sauces and dressings, in fact, at low pH and after heat treatment, this gum is considered safe to use in food materials (López-Franco et al., 2009).

The potential use of Sterculia gum as a thickening agent was investigated in printing wool, silk, and nylon-6 fabric with reactive dyes (Ibrahim et al., 2010). Natural thickening agents are high molecular weight polysaccharides such as Sterculia gum. To attain high-quality print, it is important to improve the thickening efficiency of such natural gums. This study was able to report improved thickening performance of Sterculia gum in textile printing compared to more popular guar gum.

In the paper industry, Sterculia gum is used in the production of long-fibred, lightweight papers. De-acetylated Sterculia gum is used as a binding agent in the production of long-fiber, lightweight papers (Whistler, 1993). The main purpose of Sterculia gum is to effectively prevent the fibers from forming flocks and keep them homogeneously distributed. This results in a lightweight sheet of improved formation and strength. The gum is de-acetylated in order to expose more active carboxyl and hydroxyl groups and facilitate the association with the cellulose fibers.

Silicon carbide nanoparticle incorporated Sterculia gum was studied in another work as an adsorbent for removal of cationic dyes from aqueous solution (Mittal et al., 2016). Gum polysaccharide-based hydrogels have attracted attention in the removal of dyes due to their advantageous properties like low toxicity, easy availability, low cost, and environmentally friendly nature (Mittal et al., 2013). While natural gum based hydrogels have shown significant dye removal capacity, incorporation of nanoparticles have not only improved the structural strength of the hydrogels, but also increased adsorption rate and capacity (Ghorai et al., 2014). This can be attributed to nanoparticles' ability to increase surface area to volume ratio and bring in additional adsorption sites (Mittal et al., 2014). Moreover, silicon carbide nanoparticles have higher mechanical, thermal, and electrical properties in comparison with similar nanoparticles (Zhou et al., 2006). A number of adsorbents have failed to show the acceptable result in removing cationic dyes such as rhodamine B (RhB B). But silicon carbide nanoparticles containing Sterculia gum grafted nanocomposite hydrogels exhibited excellent cationic dye removal capacity (Mittal et al., 2016). Sterculia gum based nanofiber composite was fabricated with poly (vinyl alcohol) (PVA) using the electrospun technique (Patra et al., 2015). Electrospinning, one of the most popular techniques to produce nanostructures, is carried out by dissolving the polymer in a solvent and spinning the solution to a nanometer scale under the influence of s strong electrical field (Ramakrishna et al., 2001). It is an easy and economical technique which can produce nanofibers of different properties and applications (Teo and Ramakrishna, 2006). With this technique, de-acetylated Sterculia gum and PVA were used to successfully produce composite nanofibers which exhibited increased crystallinity for application in biomedical and material science.

Sterculia gum has found application in the food industry as a stabilizer for aerated dairy products and frozen desserts in concentrations from 0.2–0.4%, this controls the formation of ice crystals in the product. But in low pH products, it is used as an acid-resistant stabilizer in sweetened drinks and fruit ices and also used in stabilizing packaged whipped cream products and meringue toppings. Moreover, this gum can prevent syneresis and improve the spreadability characteristics of cheese spreads when used in concentrations up to 0.8%. Sterculia gum is added as a binder in low-calorie dough-based products like bread, pasta, and differ bakery products. Furthermore, this gum can be used in the preparation of special quick-cooking farina cereals and ground meat products. This is because Sterculia gum has good water-holding and water-binding properties to produce good quality finished products. Lastly, Sterculia gum is also a good emulsion stabilizer for French-style salad dressings as it improves the viscosity of the aqueous phase of the oil-water emulsion (López-Franco et al., 2009).

7.7 FUTURE SCOPE

The advantages of using natural gums over synthetic in the pharmaceutical aspect are many. Natural gum such as sterculia, for example, is biocompatible and cheap. They are abundant and poses less risk. Hence the use of them, especially in pharmaceuticals, are increasing in large numbers day by day.

Natural gums have numerous usages in pharmaceutical industries, as discussed in the previous sections. They show good performance as a superdisintegrants, in the sustainable and controllable drug delivery, as a dissolution enhancer and as a drug retardant. They can be blended with other compounds to perform many other versatile actions such as in the production of hydrogel for wound dressing. By modifying sterculia gum, the shortfalls can be overcome. They can even be used as in the preparation of microspheres for anti-cancer drug delivery. In a study, microspheres made from hydrolyzed polyacrylamide graft karaya gum was used as a colon-specific drug carrier for the anti-cancer agent capecitabine (Figure 7.3). After crosslinking, the copolymer of PAAm-g-GK with glutaraldehyde (GA), the drug release amount was maximum in the colonic region showing a potential for a colon targeting characteristics (Alange et al., 2017).

It is inevitable that we are to switch to a more natural way of living for our sustainable existence. It is safe to say that the number of researches on natural gums and exploiting their potential characteristics will continue to grow.

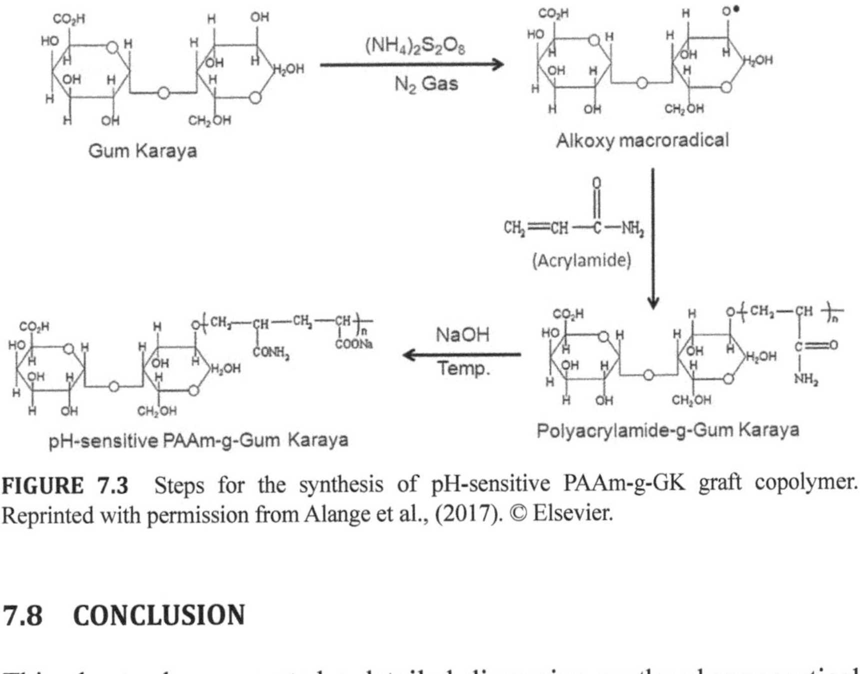

FIGURE 7.3 Steps for the synthesis of pH-sensitive PAAm-g-GK graft copolymer. Reprinted with permission from Alange et al., (2017). © Elsevier.

7.8 CONCLUSION

This chapter has presented a detailed discussion on the pharmaceutical use of sterculia gum. Sterculia gum, also known as karaya gum, has the potential to play a major role in the pharmaceutical industry. It is inexpensive, easily available and most importantly biocompatible and non-toxic. Sterculia gum can be put into diversified uses. Throughout this chapter, usage, and advantages of the use of it have been discussed. There is enough scientific evidence that karaya gum can be an exceptional ingredient in pharmaceutical aspects. More researches should be done to exploit the potential further on karaya gum.

KEYWORDS

- **drug release**
- **gum karaya**
- **pharmaceutical applications**
- **sterculia**

REFERENCES

Afrasim, M., & Shivakumar, H., (2010). Formulation and in vitro evaluation of sustained-release tablet of diltiazem: Influence of hydrophilic gums blends. *J. Pharm. Res., 3*(3), 600–604.

Alange, V. V., Birajdar, R. P., & Kulkarni, R. V., (2017). Functionally modified polyacrylamide-graft-gum karaya pH-sensitive spray dried microspheres for colon targeting of an anti-cancer drug. *Int. J. Biol. Macromol., 102*, 829–839.

Arora, G., Maliki, K., Sharma, J., & Nagpal, M., (2011). Preparation and evaluation of solid dispersions of modified gum karaya and aceclofenac: Controlled release application. *Pelagia Res. Lib.*, 142–151.

Babu, G. M. M., Prasad, C. D., & Murthy, K. R., (2002). Evaluation of modified gum karaya as carrier for the dissolution enhancement of poorly water-soluble drug nimodipine. *Int. J. Pharm., 234*(1/2), 1–17.

Bart, B. J., Biglow, J., Vance, J. C., & Neveaux, J. L., (1989). Salicylic acid in karaya gum patch as a treatment for verruca vulgaris. *J. Am. Acad. Dermatol, 20*(1), 74–76.

Biswas, G., Chakraborty, S., Majee, S., & Das, U., (2014). Insight into the release kinetics of amoxicillin trihydrate from buccoadhesive tablets with a natural gum. *Res. J. Pharm., Biol. Chem. Sci., 5*(3), 772–785.

Casadei, E., & Chikamai, B., (2010). Gums, resins and waxes. In: Singh B.P., (eds.), *Industrial Crops and Uses* (vol, 19, pp. 411–430). CABI, Wallingford, UK.

Chivate, A. A., Poddar, S. S., Abdul, S., & Savant, G., (2008). Evaluation of Sterculia foetida gum as controlled release excipient. *AAPS Pharm. Sci. Tech., 9*(1), 197–204.

Eastwood, M., Brydon, W., & Anderson, D., (1983). The effects of dietary gum karaya (Sterculia) in man. *Toxicol. Lett., 17*(1/2), 159–166.

Galla, N. R., & Dubasi, G. R., (2010). Chemical and functional characterization of Gum karaya (Sterculia urens L.) seed meal. *Food Hydrocoll., 24*(5), 479–485.

Ghorai, S., Sarkar, A., Raoufi, M., Panda, A. B., Schönherr, H., & Pal, S., (2014). Enhanced removal of methylene blue and methyl violet dyes from aqueous solution using a nanocomposite of hydrolyzed polyacrylamide grafted xanthan gum and incorporatednano-silicaa. *ACS Appl. Mater. Interfaces, 6*(7), 4766–4777.

Harshal, A., & Priscilla, M., (2011). Development and evaluation of herbal laxative granules. *J. Chem. Pharm. Res., 3*(3), 646–650.

Hirst, E. L., & Dunstan, S., (1953). The structure of karaya gum (Cochlospermum gossypium). *J. Chem. Soc. (Resumed)*, 0, 2332–2337.

Ibrahim, N., Abo-Shosha, M., Allam, E., & El-Zairy, E., (2010). New thickening agents based on tamarind seed gum and karaya gum polysaccharides. *Carbohydr. Poly.*, *81*(2), 402–408.

Imeson, A. P., (2012). *Thickening and Gelling Agents for Food*. Springer Science & Business Media.

Jani, G. K., Shah, D. P., Prajapati, V. D., & Jain, V. C., (2009). Gums and mucilages: Versatile excipients for pharmaceutical formulations. *Asian J. Pharm. Sci.*, *4*(5), 309–323.

Kalyani, V., Kishore, V. S., Kartheek, U., Teja, P. R., & Vinay, V., (2014).Comparativee evaluation of olanzapine immediate release tablets by using natural super disintegrants. *Int. J. Pharm. Res. Rev.*, *3*(4), 50–56.

Krishna, R. R., Murthy, T. E. G. K., & Himabindu, V., (2009). Design and development of mucoadhesive microcapsules of Glipizide formulated with gum karaya. *J. Pharm. Res.*, *2*(2), 208–214.

Le Cerf, D., Irinei, F., & Muller, G., (1990). Solution properties of gum exudates from Sterculia urens (karaya gum). *Carbohydr. Polym.*, *13*(4), 375–386.

López-Franco, Y., Higuera-Ciapara, I., Goycoolea, F. M., & Wang, W., (2009). 18 - Other exudates:Tragacanthh, karaya, mesquite gum andLarchwoodd arabinogalactan. In: *Handbook of Hydrocolloids* (2nd, pp. 495–534). Woodhead Publishing.

Mbuna, J. J., & Mhinzi, G. S., (2003). Evaluation of gum exudates from three selected plant species from Tanzania for food and pharmaceutical applications. *J. Sci. Food Agric.*, *83*(2), 142–146.

Meier, P., Seiler, W., & Stähelin, H., (1990). Bulk-forming agents as laxatives in geriatric patients. *Schweiz. Med. Wochenschr*, *120*(9), 314–317.

Mills, P. L., & Kokini, J. L., (1984). Comparison of steady shear and dynamic viscoelastic properties of guar and karaya gums. *J. Food Sci.*, *49*(1), 1–4.

Mittal, H., Maity, A., & Ray, S. S., (2015). Synthesis of co-polymer-grafted gum karaya and silica hybridorganic-inorganicc hydrogel nanocomposite for the highly effective removal of methylene blue. *Chem. Eng. J.*, *279*, 166–179.

Mittal, H., Maity, A., & Ray, S. S., (2016). Gum karaya based hydrogel nanocomposites for the effective removal of cationic dyes from aqueous solutions. *Appl. Surf. Sci.*, *364*, 917–930.

Mittal, H., Mishra, S. B., Mishra, A., Kaith, B., Jindal, R., & Kalia, S., (2013). Preparation of poly (acrylamide-co-acrylic acid)-grafted gum and its flocculation and biodegradation studies. *Carbohydr. Polym.*, *98*(1), 397–404.

Mittal, H., Parashar, V., Mishra, S., & Mishra, A., (2014). Fe_3O_4 MNPs and gum xanthan based hydrogels nanocomposites for the efficient capture of malachite green from aqueous solution. *Chem. Eng. J.*, *255*, 471–482.

Mortensen, A., Aguilar, F., Crebelli, R., Di Domenico, A., Frutos, M. J., Galtier, P., Gott, D., Gundert-Remy, U., Lambré, C., Leblanc, J. C., Lindtner, O., Moldeus, P., Mosesso, P., Oskarsson, A., Parent-Massin, D., Stankovic, I., Waalkens-Berendsen, I., Woutersen, R. A., Wright, M., Younes, M., Brimer, L., Peters, P., Wiesner, J., Christodoulidou, A., Lodi, F., Tard, A. & Dusemund, B., (2016). Scientific opinion on the re-evaluation of karaya gum (E 416) as a food additive. *EFSA Journal*, *14*(12), *4598*, 44 pp. doi: 10.2903/j.efsa.2016.4598.

Munday, D. L., & Cox, P. J., (2000). Compressed xanthan and karaya gum matrices: Hydration, erosion and drug release mechanisms. *Int. J. Pharm.*, *203*(1/2), 179–192.

Nagpal, M., Rajera, R., Nagpal, K., Rakha, P., Singh, S., & Mishra, D., (2012). Dissolution enhancement of glimepiride using modified gum karaya as a carrier. *Int. J. Pharm. Invetig.*, *2*(1), 42.

Nath, B., & Nath, L. K., (2012). Design, development, and optimization of sterculia gum-based tablet coated with chitosanEeudragit RLPO mixed blend polymers for possible colonic drug delivery. *J. Pharm.*, 2013.

Nath, B., & Nath, L. K., (2013). Studies onsterculiaa gum formulations in the form of osmotic core tablet for colon-specific drug delivery of azathioprine. *PDA J. Pharm. Sci. Technol.*, *67*(2), 172–184.

Nussinovitch, A., (1997). *Hydrocolloid Applications: Gum Technology in the Food and Other Industries*. Springer.

Park, C. R., & Munday, D. L., (2004). Evaluation of selected polysaccharide excipients in buccoadhesive tablets for sustained release of nicotine. Drug Dev. Ind. Pharm., 30(6), 609–617.

Patra, N., Martinová, L., Stuchlik, M., & Černík, M., (2015). Structure-propertyy relationships in Sterculia urens/polyvinyl alcohol electrospun composite nanofibers. *Carbohydr. Polym.*, *120*, 69–73.

Phillips, G., & Williams, P., (2001). Tree exudate gums: Natural and versatile food additives and ingredients. *Food Ingredients Anal. Int.*, 26.

Prasanthi, N., Manikiran, S., & Rao, N. R., (2011).In-vitroo drug release studies of Ziprasidone from tablets using natural gums from biosphere. *Schol. Res. Lib.*, *3*(2), 513–519.

Ramakrishna, S., Mayer, J., Wintermantel, E., & Leong, K. W., (2001). Biomedical applications of polymer-composite materials: A review. *Compos. Sci. Technol.*, *61*(9), 1189–1224.

Rana, V., Rai, P., Tiwary, A. K., Singh, R. S., Kennedy, J. F., & Knill, C. J., (2011). Modified gums: Approaches and applications in drug delivery. *Carbohydr. Polym.*, *83*(3), 1031–1047.

Reddy, M. M., Reddy, J. D., Moin, A., & Shivakumar, H., (2012). Formulation of sustained-release matrix tablets using cross-linked karaya gum. *Trop. J. Pharm. Res.*, *11*(1), 28–35.

Rowe, R., (2009). In: Raymond, C. R., Paul, J. S., & Marian, E. Q., (eds.), *Handbook of Pharmaceutical Excipients*. Chicago, APhA. Pharmaceutical Press.

Singh, B., & Pal, L., (2012). Sterculia crosslinked PVA and PVA-poly (AAm) hydrogel wound dressings for slow drug delivery: Mechanical, mucoadhesive, biocompatible and permeability properties. *J. Mech. Behav. Biomed. Mater.*, *9*, 9–21.

Singh, B., & Sharma, N., (2008). Development of novel hydrogels by functionalization of sterculia gum for use in anti-ulcer drug delivery. *Carbohydr. polym.*, *74*(3), 489–497.

Singh, B., Sharma, V., & Pal, L., (2011). Formation of sterculia polysaccharide networks by gamma rays induced graft copolymerization for biomedical applications. *Carbohydr. Polym.*, *86*(3), 1371–1380.

Sparberg, M., Van Prohaska, J., & Kirsner, J. B., (1966). Solid state karaya gum ring for use in disposable and permanent ileostomy appliances. *Am. J. Surg.*, *111*(4), 610–611.

Steinhardt, A., & Goldwater, F., (1962). Gelatin adhesive pharmaceutical preparations. *US Patent*, *3*(29), 187.

Sukhavasi, S., & Kishore, V. S., (2012). Formulation and evaluation of fast dissolving tablets of amlodipine besylate by using hibiscus rosa-sinensis mucilage and modified gum karaya. *Int. J. Pharm. Sci. Res., 3*(10), 3975.

Tekade, B. W., & Chaudhari, Y., (2013). Gums and mucilages: Excipients for modified drug delivery system. *J. Adv. Pharm. Edu. & Res, 3*(4), 359–367.

Teo, W. E., & Ramakrishna, S., (2006). A review on electrospinning design andnanofibrer assemblies. *Nanotechnol., 17*(14), R89.

Verbeken, D., Dierckx, S., & Dewettinck, K., (2003). Exudate gums: Occurrence, production, and applications. *Appl. Microbiol. Biotechnol., 63*(1), 10–21.

Whistler, R. L., (1993). Exudate gums. In: *Industrial Gums* (3rd edn., pp. 309–339). Elsevier.

Wilson, M., & Harvey, W., (1989). Prevention of bacterial adhesion to denture acrylic. *J. Dent., 17*(4), 166–170.

Zhou, W., Liu, X., & Zhang, Y., (2006). Simple approach to βSiCC nanowires: Synthesis, optical, and electrical properties. *Appl. Phys. Lett., 89*(22), 223124.

CHAPTER 8

Pharmaceutical Applications of Okra Gum

SHANTA BISWAS, SADIA SHARMEEN, MD MINHAJUL ISLAM,
MOHAMMED MIZANUR RAHMAN, PAPIA HAQUE, and
ABUL K. MALLIK

*Department of Applied Chemistry and Chemical Engineering,
Faculty of Engineering and Technology, University of Dhaka,
Dhaka-1000, Bangladesh*

ABSTRACT

In general, gums having a complex, branched polymeric structure, show high cohesive and adhesive properties. Such properties are particularly desirable in pharmaceutical preparations. Okra gum is collected from plant-derived Okra fruit (*Hibiscus esculentus* (family Malvaceae)), which is a natural polysaccharide that consists of d-galactose, l-rhamnose, and l-galacturonic acid. The plant is capable of tolerating heat and drought as well as the fruit is very popular mainly in Asia and Africa, which has opened the door to consider Okra other than food. These exorcisms of Okra gum lead to aptness as excipients in the development of different pharmaceutical formulations. In continuation, in many literatures, as a pharmaceutical excipient, the efficiency of Okra gum has received considerable attention as a binder to tablets giving good hardness and friability, controlled release agent, film coating material, bio-adhesive, suspending agent, sustained-release drug delivery matrices, disintegrate, etc. In addition, comparing to commercial synthetic polymers, it has vast advantages, e.g., it is safe, biodegradable, biocompatible, chemically inert, non-irritant, eco-friendly, and biocompatible. One of the most common uses of Okra gum is in Paracetamol tablet providing controlled release after intake. In this chapter, we have tried to illustrate the significant pharmaceutical applications of Okra gum as well as to enlighten the potential implication on future natural gum applications.

8.1 INTRODUCTION

Due to the recent development in natural products, researchers are tending towards the replacement of synthetic polymers by various natural materials. Among these natural materials, plants have played a vital role by supplying some useful polymers through their leaves, seeds, barks, gums, mucilage, and so on. Nowadays, large numbers of pharmaceutical excipients are based on plants that are readily available. Gums are most frequently observing plant materials, usually refer to polysaccharide hydrocolloids, which do not comprise a part of the cell wall (Gt et al., 2002) but are exudates with a wide range of pharmaceutical product based applications. It is composed of 'polyuronides' (calcium, potassium, and magnesium salts of complex substances). The gums are frequently used as a thickening, binding, disintegrants, gelling agents, laxative agents, emulsifying, suspending, and stabilizing agents in pharmaceutical industries. They have also been applied as matrices for sustained release of drugs. These hums are applied due to their high abundance in nature, safety, economy, biocompatibility, and low cost. Moreover, natural gums have advantages over synthetic ones since they are non-toxic, chemically inert, low cost, biodegradable, and most common in nature. Several kinds of modifications are also possible in these gums to have tailor-made materials for drug delivery systems and thus can compete with the available synthetic excipients (Prajapati et al., 2013). In terms of structure natural gums form three-dimensional monomeric networks, which help in trapping water, drug, and other excipients in the vicinity. But these natural gums have certain disadvantages such as; susceptibility to microbial attack because of its moisture content, batch to batch variation due to geographical and environmental effect, uncontrolled rate of hydration and decreasing in viscosity upon storage (Rowe et al., 2006; Bhosale et al., 2015). Guar gum, xanthan gum, locust bean gum (LBG), ghatti gum, cashew gum, tamarind seed gum, karaya gum, mango gum, and gellan gum are some examples of natural gums.

A water-soluble derivative of Okra plant (*Hibiscus esculentus L.*) is Okra gum. It is the most common gum used in cooking purpose at home mostly as a thickener in soups (Ndjouenkeu et al., 1997). Its common name is Okra and also renowned as lady's fingers, bamia, gombo, etc.

In the case of pharmaceutical applications, it has been analyzed as a binding agent for tablets. And it has also been experienced to make tablets with satisfactory friability, good hardness, and drug release profiles (Okoye et al., 2011).

In a study, as a binder of paracetamol tablet okra gum has been evaluated and it was found to show a quicker onset and higher quantity of plastic deformation than those tablets containing gelatin (Avachat et al., 2011). Although a medicine containing Okra gum results in higher disintegration time. Okra gum, when extracted in water, can produce a remarkably viscous solution with a slimy appearance. This highly viscous property may proof Okra gum as an efficient retarding polymer in the formulation of sustained-release tablets. Concerning on many studies in recent time it is seen that okra gum has been proved as a beneficial hydrophilic matrixing agent in sustained drug delivery devices. Besides sustained released tablet, Okra gum is considered as a promising agent for controlled drug delivery system. It shows a better result than the other possible available natural polymeric matrices (Avachat et al., 2011). In addition, Okra gum is widely applied in manufacturing many foodstuffs like chocolate, cookies (Romanchik-Cerpovicz et al., 2002a).

In many scientific research papers, Okra gum is chosen as a possible agent for various pharmaceutical agents (Newton et al., 2015; Sinha et al., 2015a, b). This chapter focuses on a brief description of the recent development of Okra gum in the pharmaceutical field. The first part of the chapter will summarize the sources, characteristics, method of extraction, and basic applications of this gum. The next part will emphasize on the recent development of Okra gum as a pharmaceutical coating agent, binding agent, disintegrant, bio-adhesive, material for controlled and sustained release of the selective drug, and so on.

8.2 OKRA GUM

8.2.1 SOURCES

Okra gum, a natural polymer, is extracted from a tall erect annual plant scientifically known as *Abelmoschus esculentus*. Its common name is Okra or lady's finger and popularly known as *bhindi* in India and as *bamies* in Mediterranean and Arab countries. Okra was previously considered in *hibiscus* family, but now it belongs to distinct genus *Abelmoschus* in

Malvaceae family. This plant is annually cultivated in the tropical and subtropical regions, particularly in Africa, Asia, and Northern Australia and consumed as a vegetable. Okra is a fast growing plant and can be grown in almost all soil types. Moreover, this is one of the major heat and drought-tolerant plants (Bakre and Jaiyeoba, 2009b).

8.2.2   METHOD OF EXTRACTION

Fresh okra pods were collected, washed, and sliced into two halves. Seeds were separated before extraction. The prepared okra halves (100 g) were blended in 500 mL 0.05 M NaOH for 5 minutes. For this blending purpose, a heavy-duty blender is needed. After that, the mixture is needed to be centrifuged at 1×2000 g. The supernatant was separated, and the extraction was repeated upon precipitation. After adjusting the pH of combined supernatant at fixed 7, it was freeze-dried, ground, and stored at 4°C in an airtight glass container for future use (Alamri et al., 2012).

8.2.3   PHYSICAL AND CHEMICAL PROPERTIES

Okra gum is random structurally made of polysaccharides coil composed of rhamnose, galactose, and galacturonic acid. The repeating units of the gum were observed to be (1-2)-rhamnose and (1-4)-galacturonic acid residues with disaccharide side chains and a degree of acetylation (DA = 58) (Alamri et al., 2012) (Figure 8.1). Chemically this gum is inert, nonirritant, biodegradable, and biocompatible (Zaharuddin et al., 2014).

Okra gum in powder form showed low solubility in water but insolubility in other solvents like acetone, chloroform, and ethanol. The sparingly soluble nature of okra gum in water helped to create viscous dispersion, which can be useful in strong polymeric matrix system in controlled release of drugs. Moreover, the insolubility of okra gum in acetone helped to create dried okra gum, where acetone acts as good precipitating and drying agent (Zaharuddin et al., 2014).

Okra gum exhibited higher viscosity with increasing concentration. The higher viscosity of gum tends to increases stickiness, which results in a higher possibility of holding ingredients. This attributed can be used in producing tablets with slower drug release property. The viscosity of 1% solution of Okra gum was found to be 228.78 cP, whereas the viscosity of 0.5% solution of Okra gum showed only 62.32 cP (Zaharuddin et al., 2014; Kalu et al., 2007).

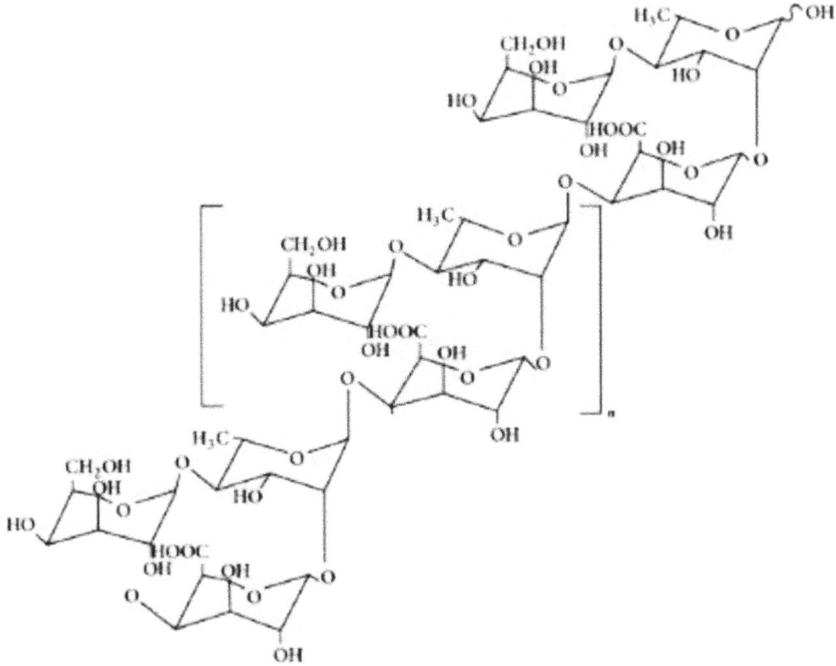

FIGURE 8.1 Chemical structure of polysaccharide of Okra gum. Reprinted from Zaharuddin et al., 2014. Open access.

It is previously stated that polysaccharides from Okra are soluble only in alkaline or acidic solutions. Besides these, it is also soluble in aqueous alcoholic solution. When extracted in water, these polysaccharides can result in a highly viscous solution with slimy appearance. At pH 4–6, the viscosity of Okra extract in sodium borohydride exhibited a pseudo-plastic behavior. In case of determining the trends in viscosity, it was observed that by adding water solubles, the viscosity of Okra extract decreased, and it can be increased by adding maltodextrins. The Okra extract was also found to strengthen weak dough made from soft wheat flour (Bhat and Tharanathan, 1987).

The pH of Okra gum was observed to be 6.59. Okra gum exhibited maximum viscosity at neutral pH. This facilitates the retarding capacity of sustained-release tablets, where neutral pH is of great significance as it keeps irritation to the gastrointestinal tract (GIT) to a minimum (Malviya).

Okra gum is a polysaccharide that can attach water molecules via hydrogen bond. This is the reason behind the relatively high moisture content of 14.83% found in Okra gum. This bound moisture has notable effects on tablet formation. By the formation of moisture film on the particles, it can affect the compressibility of tablets. This, in fact, works as lubrication and facilities easy flow of the powder during the tablet manufacturing process. This moisture film on the Okra powder also makes the ingredients stick with each other resulting in more intact tablet (Nep and Conway, 2010; Bakre and Jaiyeoba, 2009b). Okra gum consists of amorphous and crystalline structure, which is visible from the X-ray diffraction pattern. For this reason, Okra gum exhibits both glass transition temperature (T_g) and melting point (T_m) at 60°C and 180°C, respectively (Zaharuddin et al., 2014). Below glass transition temperature, the gum stays at glassy state as the molecules are not able to move around freely. At the glass transition temperature, the molecules start to vibrate. Whereas above glass transition temperature, due to high molecular mobility, okra gum is in the rubbery state (Jadhav et al., 2009).

8.3   GENERAL APPLICATIONS OF OKRA GUM

Okra gum has been used in pharmaceutical applications, mainly as, binder, control release, film coating, bio-adhesive, suspending agent and other applications such as thickener in soups and fat ingredient substitute.

Okra gum was used as a tablet binder by Tavakoli et al. (2008) and Emeje et al. (2007). It was possible to fabricate tablet with okra gum as a binder and investigate drug release capacity of acetaminophen, ibuprofen, and calcium acetate tablets. The tablet formulation exhibited good friability, hardness, and disintegration time and dissolution rate. However, using okra gum as a binder prolonged dissolution rate of some slightly soluble drug. This led to the conclusion that okra gum could be a good candidate for sustained release formulation only.

The applicability of okra gum in controlled drug release was investigated by Newton et al. (2015). Propranolol HCl delayed release matrix tablets were formulated using Okra gum. The main purpose was to develop a colon targeted drug delivery system to treat early morning sign in blood pressure. Details of this study have been added in section

8.4.1 of this chapter. In another work, film coating efficacy of okra gum was investigated using acetaminophen as a drug (Ogaji and Nnoli). Film coating is a unit operation used to improve stability, mask unpleasant taste, and odor and modify release characteristics of the drug (Rhodes and Porter, 1998; Kwok et al., 2004). Polymeric solution or dispersion has been a popular candidate for film coating formulation. On that basis, okra gum, a natural polymer, was studied for its applicability as a film coating material. The physiochemical properties of Okra gum formulation did not differ substantially compared to popular film coating material such as hydroxypropyl methylcellulose (HPMC). The results led to the deduction that Okra gum can be used as a film-coating ingredient.

Okra gum has also been used in non-pharmaceutical application such as fat ingredient substitute. Reducing dietary fat intake in a product such as ice cream, chocolates, and other dairy dessert has gained traction recently. This is where Okra gum has become a prominent component. Okra gum was investigated as a low-cost substitute for milk fat in a chocolate frozen dairy dessert such as ice cream (Romanchik-Cerpovicz et al., 2006b). Okra gum was found to be an acceptable replacement; moreover, it was capable of imparting many acceptable sensory qualities of milk fat. A thorough study on textural, flavor, and aftertaste property of Okra gum might reveal its feasibility in other food products like salad dressing or sauce.

Besides that, traditionally, Okra is used to thicken soups, and the forms may be fresh or dried and ground. Sometimes this crushed dried Okra is used as a flavoring to the soups. A very interesting use of Okra gum from the non-pharmaceutical background is the use of immature fruit of the Okra in folk medicine. It was applied as a diuretic and for the treatment of dental disease (Bhat and Tharanathan, 1987). Moreover, like egg white substitute, polysaccharides obtained from okra are used (Costandino and Romanchick-Cerpovicz, 2004).

8.4 PHARMACEUTICAL APPLICATION OF OKRA GUM

8.4.1 CONTROLLED AND SUSTAINED RELEASE OF THE DRUG

All pharmaceutical dosage forms consist of thousands of excipients along with the active ingredients to assist manufacturing technique. Gums and mucilages obtained from the gum can comply with many requirements of pharmaceutical excipients because they are non-toxic, inexpensive, readily available, stable, associated with less regulatory issues as compared to their synthetic counterpart. Moreover, modification can be performed to have the desired effect. Natural gums and mucilage are composed of many constituents. Among them, the polysaccharides, resins or the tannins present in the gum are held responsible for having a release retardant properties to the dosage form. Gums are obtained from different parts, e.g., the epidermis of the seed while on the other hand, it may be extracted from the leaf or bark. Okra gum based drug release-retarding materials has been recently studied both as carriers in conventional sustained release dosage forms and in buccal, gastroretentive, and microcapsules based systems.

In a study, in modified release matrices, Okra gum was studied as a controlled-release agent. Controlled release of paracetamol was found for more than 6 hours by using Okra gum. Time-Independent kinetics are observed in case of release rates of the drug. Sodium carboxymethyl cellulose (NaCMC) and hydroxypropylmethylcellulose (HPMC) are tested as a support material for sustained-release tablet and then compared to Okra gum. It was observed that Okra gum supported better than the NaCMC and HPMC (Kalu et al., 2007). It was also found that when combining Okra gum matrix and NaCMC, or on further addition of HPMC showed in near zero order release of paracetamol. This resulted that for a long period of up to 6 hours, Okra gum matrices could be beneficial in the formulation of sustained-release tablets. Noted that, the rate of drug release was governed by the concentration of the drug component present in the prepared matrix (Avachat et al., 2011).

In a different study, a chronotherapeutic drug delivery system was designed by using various natural polymers. It was developed to treat early morning sign in blood pressure for colon targeting drug delivery systems. In this study, polymers such as Tamarind gum (TmG), Okra gum, and Chitosan (CS) were used. Natural polymers are better than the synthetic polymer with respect to their well-oriented molecular structure. This property boosts up their strength and biocompatibility (Newton et al., 2015). At the present time, most of the drugs that are available for colon-targeted action are based on the specific pH sensitivity. This kind of drug does not start functioning unless it is exposed to its desired pH-bearing atmosphere. But, these factors are sometimes very difficult to maintain in case of synthetic polymer. Whereas, natural polymers are having attractive properties are preferred for this kind of drug designing. Here, in this case, Propranolol HCl was incorporated in the formulation as a model drug. After incorporation of the drug, the order for assessing the controlled release system and time-dependent release potential of various natural polymers were measured as primary polymer TmG was extracted and used. In some cases, an auxiliary polymer, Carbopol 940 was used to modify the drug release and physicochemical nature of the prepared tablets. The evaluation report of this Okra gum based tablet is very satisfactory. The in vitro analysis was performed under comparable physiological condition. The release profile was performed for 1.5 h in 0.1N HCl, followed by a phosphate buffer of pH 6.8 for 2 h and pH 7.4 phosphate buffers till the highest amount of drug release. This analysis confirmed the possibility of using these polymers in this selective purpose (Newton et al., 2015).

A different study was conducted by Zaharuddin et al., (2014) that evaluated the efficiency of Okra gum in sustaining the release of propranolol hydrochloride in a tablet. The main ingredient Okra gum was extracted from the pods of *Hibiscus esculentus* using acetone as a drying agent. After drying the Okra, gum is powdered. Several physical and chemical properties such as pH, moisture content, solubility, viscosity, morphology study, crystallinity, and thermal behavior were necessary to determine for its application as a drug excipient. By following granulation and compression methods, the gum powder was used in the fabrication of tablet. As a model drug, propranolol hydrochloride was used. To prove the better performance of Okra gum for this selective application, it was compared with some well-known synthetic and semi-synthetic binders, which are sodium alginate (SA) and HPMC, respectively. The analysis showed that the Okra gum retarded the release of the drug for up to 24 hours. It was also observed that it possessed the longest release profile compared to SA and HPMC. As a part of the evaluation, by conducting hardness and friability tests, the tensile and crushing strength of tablets was also evaluated. Tablets based on Okea gum resulted in the highest hardness value and lowest friability. Hence, Okra gum was testified as a salutary to produce tablets with favorable sustained release phenomena, strong tensile, and better crushing strength (Zaharuddin et al., 2014) property.

Attama et al., reported that in the formulation of mucoadhesive indomethacin (IND) tablets, Carbopol 941 and *Abelmoschus esculentus* gum (Okra gum) were applied as bioadhesive polymers. Okra based tablets were compared with a tablet coated with 50% w/v solution of Eudragit I. 100 in ethanol. Tablet hardness, uniformity of weight, disintegration time, friability, and absolute drug content, etc., were evaluated as a parameter of physical properties of a tablet. Moreover, the release data of the selected drug were measured by using a simulated intestinal fluid (SIF pH 7.2) without pancreatin, and in 0.1 N solution of HCl. Results obtained from physical properties and release profile. It was observed that the tablet with an equal ratio of 941 and the best bioadhesive strength characteristics for both coated and uncoated tablets was obtained by using Okra gum in a particular ratio (1:1). The drug-release percentage of drug material was observed are 53–90% for uncoated tablets in 0.1 N HCl and SIF and whereas a value of 9–16% was found for coated tablets in 0.1 N HCl. In the case of coated tablets in SIF, it was found 63–100% after 8 hours (Attama et al., 2003).

A study revealed the successful fabrication of Zinc (Zn^{2+}) ion induced diclofenac sodium (DS)-loaded alginate-okra (*Hibiscus esculentus*) gum (OG) blend beads by Sinha et al. The beads were fabricated in a complete aqueous environment through Zn^{2+} ion induced ionic-gelation cross-linking method. The optimized formulation of Zn^{2+} ion induced DS-loaded alginate-OG beads resulted in 89.27 ± 3.58% of drug encapsulation efficiency. Formulated bead sizes were obtained within 1.10 ± 0.07 to 1.38 ± 0.14 mm. The drug-polymer interaction, bead surface morphology in the optimized bead matrix was analyzed for better understanding of the chemistry. Over a prolonged period of 8 hours, these formulated beads showed a sustained *in vitro* drug release. A controlled-release (zero-order) pattern with super case-II transport mechanism was followed. The swelling and degradation parameter of the beads were also observed, and it was found by the researchers that these two properties were affected by the pH of test mediums. It is noted that adjusted pH is essential for intestinal drug delivery (Sinha et al., 2015b).

Sinha et al., studied the efficacy of isolated OG. It was evaluated as a promising sustained drug release ingredient. For oral use, it is combined to SA (resulted in a polymer-blend) in the development of controlled glibenclamide release ionically-gelled beads. It was previously stated that Okra gum was isolated from Okra fruits. The solubility, pH, viscosity, and moisture content of the gum were studied as it was done in previous cases. This kind of analysis is very important for the practical application of this gum. Concentrated on the preparation, ionic-gelation method was applied through a cross-linker. Here, Glibenclamide-loaded OG-alginate gelled blend beads were fabricated where $CaCl_2$ is used as a cross-linking agent showing a drug entrapment efficiency of 64.19 ± 2.02 to 91.86 ± 3.24% and the bead sizes varied within 1.12 ± 0.11 to 1.28 ± 0.15 mm, and finally the beads exhibited sustained *in vitro* drug release over a prolonged period of 8 hours. Here a controlled-release (zero-order) pattern with super case-II transport mechanism was followed for the *in vitro* drug release from these OG-alginate beads. The pH of the test medium highly dominates the swelling and degradation phenomenon of these beads (Sinha et al., 2015a).

8.4.2 TABLET BINDER

Tablets are renowned dosage forms prescribed worldwide because of their portability and competence. Thus, the continuous improvement of tablet formulations always remains as a concern for the pharmaceutical industries (Symecko and Rhodes, 1995). However, one of the major improvements was the use of excipients named binders or adhesives during the formulations (Horisawa et al., 1993). In the tablet formation process, binders actually provide the required adhesive force among the powder particles to form the granules, which ultimately forms compacts by the application of pressure. Commonly, the higher compact ability of the tablets resides proportionally with the strength of the binder, and on the contrary, the disintegration time of the resulted tablet will be longer. In the time of granulation polymeric binders, due to their film-forming properties impacts their distributions over the substrates imparting the major characteristics of the final tablets and granules (Adebayo and Itiola, 2003). There are many synthetic materials used as tablet binder like starch, HPMC, etc. Recently, mucilage and gums obtained from natural polymers are often preferred than synthetic products because these are non-toxic, cost-effective, and easily available (Baveja et al., 1988). A later study suggested that the okra gum as a binder is superior from povidone, gelatin, and HPMC as it has the proficiency to decrease brittle fracture tendency in paracetamol tablets. It also concluded that the manufacturing cost in tablet production with okra gum was much lower than the other sources as a binder so it can be exploited as an alternative (Okoye et al., 2011).

Okra is notable for its highly viscous mucilage at low concentration, and because of this property, it has been received attention to the researchers as a potential binder, i.e., pharmaceutical excipient. The adhesive nature helps the powder to aggregate and form the granules using cohesiveness (Onunkwo, 1996; Momoh et al., 2008).

In tablet dosage form as a binding agent, okra gum has been sought in many journals and tablets made by okra gum has been proved with good hardness, drug release properties and friability (Kalu et al., 2007). Some research has also been reported as a plasma expander (Farinde et al., 2007). So, the manufacturing companies should take advantages of this economic source of top-notch pharmaceutical excipient that has been reported (Nasipuri et al., 1997).

8.4.3   COATING AGENT

Okra gum can be act as a useful coating agent which results sustain drug release characteristics, or it can modify the drug with a good film coating for the survival of drug degradation in the stomach. The gum from okra has been reported as a potential film-coating agent of paracetamol tablets (Ogaji and Nnoli). The research envisaged that the coating of drug with okra gum showed better physicochemical properties than the core tablets and can mask the taste and odor efficiently. However, the dissolution properties ware not hampered much as per the investigation. Okra gum has been observed as a control drug-releasing agent comparing with sodium carboxymethyl cellulose and HPMC in Paracetamol tablet formulation (Kalu et al., 2007). The results observed were satisfactory as okra gum matrices effectively sustained the release of tablets more than 6 hours than the others with the time-independent releasing kinetics. The kinetics alters the dissolution profiles making it more favorable as a drug release controller.

8.4.4   BIO-ADHESIVE MATERIAL

Bio-adhesion refers to a phase of two substances should be held together for a longer time by interfacial forces where at least one should be biological in nature. Okra gum can be used as bio-adhesive materials in drug delivery system. Because of its highly viscous mucilage at low concentration, it can firmly attach other materials for a longer period. This property of okra gum has been recently taken into consideration (Attama et al., 2003).

Due to the biocompatible and biodegradable nature of okra gum, it is currently under the supervision for the development of novel drug set up.

In order to that mucoadhesive drug delivery system has been introduced. It mainly represents the controlled release drug delivery system which absorbs the drug for the required duration periods, i.e., helps the drug to sustain inside the stomach for a longer period. For this reason, drug wastages can be decreased, and bioavailability, drug solubility can be increased (Sharma et al., 2013). This type of drug delivery system is very much effective for nasal delivery applications.

Control drug release of antidiabetic drug glibenclamide has recently been provided by the researchers. Glibenclamide release fastens in the gastric juice when taken orally. The drug release pattern had been improved by forming mucoadhesive beads using isolated okra gum and SA with the drug entrapment. The sustainability can be slower over 8 hours, which causes better bioavailability and at the same time, reduce dosage intervals (Sinha et al., 2015a).

8.4.5 SUSPENDING AGENT

Okra gum can be proved as potential emulsifying and suspending agents by effectively stabilizing the emulsion. It can provide multimolecular condensed films of high tensile strength, which prevents coalescence of the droplets via interfacial absorption. Usually, it forms a hydrophilic barrier between two phases sealing all the hydrophobic particles. The hydration layer can be strengthened through hydrogen bonding and interactions between the molecules. Okra gum does not reduce any interfacial or surface tension that's why it needs a wetting agent while working as an emulsifier.

It can also be used as hydrophilic colloids, which can be used as a continuous viscous phase making the solid particles of the drug materials suspended for a long time, which confirms the uniformity of the drug dose. The emulsifying action of okra gum was listed in the literature (Ghori et al., 2014). It was evident that okra gum can act as an emulsifier in the low pH governed environment for oil in water type emulsion system with n-hexadecane as the dispersed phase.

8.4.6 PHARMACEUTICAL DISINTEGRATES

Okra gum can absorb water and swell, which governs the disintegration property of the manufactured tablet using it as an excipient. Usually, it can swell five times higher than the original volume. This swelling leads the breakage of the tablets, which increases the dissolution rate of the drug. Due to the presence of mucoadhesive properties, okra gum can also increase the residence time of the tablet by bioadhesion.

Several reports projected the idea of okra gum as pharmaceutical disintegrates where okra gum represented comparatively as a better swelling and mucoadhesive agents which ultimately enhanced the residence time due to its bulkiness and better bio-adhesion with the targeted region.

The dissolution profiles of Propranolol HCl were achieved by using the newly developed matrix using okra gum. Okra gum showed controlled release, as well as followed zero-order kinetics confirming drug release, was independent of the concentration (Newton et al., 2015).

A comparing study evaluated okra gum as a disintegrating agent with gelatin in paracetamol tablet formulations noting that plastic deformation and faster onset were higher with formulations containing okra gum than gelatin. It was also observed that the crushing strength and disintegration time of the tablet increases and friability decreases with respect to the increased okra gum concentration. Comparison concluded that tablets with gelatin produce higher crushing strength, and tablets with okra gum showed a longer disintegration time. It was suggested that in sustained drug release devices okra gum could be a useful hydrophilic matrix (Emeje et al., 2007).

Another research projected that okra gum could prolong dissolution rate of some slightly soluble drugs hence sustained the drug release. However, it was proved that okra gum provided sufficient hardness, low friability and desirable disintegration time for some tablets like Acetaminophen, Ibuprofen, and Calcium acetate with the percentage of drug release about 15%, 44%, and 96% respectively after a certain time. Another researcher reported that okra gum could formulate some tablets with good hardness and friability (Tavakoli et al., 2008). The beads of Zn^{2+}-ions induced okra gum blended with alginate also showed swelling dependent on pH, which can be applied for intestinal drug delivery (Sinha et al., 2015b).

8.4.7   OTHER MINOR APPLICATION

Okra gum recently is using as the main constituent in some commercially available products of food and medicine. Due to some potential characteristics like rheological behavior, ability to stabilize the acidic emulsion and formulation of oil-water emulsion make it an effective agent for the applications like composite materials and food foams. It is also found its application as an edible coating and as a carrier. Already some researchers concluded about the development of the nano-scale carriers using okra gum as a key ingredient for the improved drug delivery systems (Bhattacharyya, 2012; Roy et al., 2012; Ofoefule and Chukwu, 2001).

Okra gum was used as a successful release modifier which acts as a matrix sustaining the drug release of diclofenac and furosemide tablets. About only 10–15% concentration of okra gum in the matrix was capable of extending the release of drug for about 10 hours. Also, it was found that the drug release followed nearly zero order kinetics with the release mechanism of the Korsmeyer-Peppas model (Pawan and Nitin). Another present use of okra gum is as a carrier for other drug deliveries. Okra gum was used in metronidazole tablet formulation as a medium as per the report by Bakre and Jaiyeoba (2009a).

However, as carrier okra gum is mainly potential for oral gel and nasal spray due to its mucoadhesive properties. Okra gum with polymer blend formed gelling beads due to the bio-adhesive properties, which can be potential in oral use. It was used in developing for a sustainable carrier for the ironical release of glibenclamide. Mucoadhesive beads are also finding their applications in other areas as they can be produced to encapsulate other drugs that need a sustained release for better bioavailability and thus can reduce the dose interval.

Another research developed mucoadhesive gel using okra gum with the drug rizatriptan benzoate for the nasal delivery system (Table 8.1). The same research group also prepared and evaluated the mucoadhesive microspheres using okra gum as a novel carrier for safe and effective delivery of rizatriptan benzoate into nasal delivery (Chanchal et al., 2018).

8.5 FUTURE ASPECT OF OKRA GUM IN PHARMACEUTICAL APPLICATIONS

Okra plant is found in almost all regions of the world, especially in South Asian, Ethiopian, and West African regions. From the previous sections of this chapter, it is found that it is very easy to extract with the least number of chemical and equipments. It is widely used both in food and non-food purposes. Considering pharmaceutical background, very few drugs have been explored in combining with Okra gum for various applications. Being a non-toxic, less expensive and highly available material, it is expected to explore the probable use Okra gum in variegated fields of pharmaceutical application like a chemotherapeutic agent, tissue engineering, base of anti-fungal ointments, skin care products and so on.

TABLE 8.1   Minor Applications of Okra Gum

SL. No.	Function	Reference
1.	Milk-fat ingredient substitute in chocolate frozen dairy dessert	Romanchik-Cerpovicz et al., 2006a
2.	Okra gum as a fat ingredient substitute (moisture retention in chocolate bar cookies)	Romanchik-Cerpovicz et al., 2002b
3.	Physicochemical properties of acetaminophen pediatric suspensions formulated with Okra gums obtained from different extraction processes as a suspending agent	Ikoni, 2014
4.	Low-fat banana bread by using Okra gum as a fat replacer	Hu and Lai, 2017
5.	Okra gum on pasting and rheological properties of cake-batter	Qasem et al., 2017
6.	Okra gum fortified bread	Alamri, et al., 2012

8.6   CONCLUSION

Application of natural gums for pharmaceutical purpose is always very lucrative because they are numerously available, potentially economical, non-toxic, and their ease of chemical modifications without showing any adverse effect have made their use so attracting. Almost all of the gums are potentially biodegradable and with few exceptions, and also biocompatible. Okra gum possesses almost all the properties of a typical natural gum. Over the years, researchers have developed a good variety of composites based on Okra gum. A very few works have been observed based on modified Okra gum. Because of their well controllable properties, they become one of the most attractive materials for numerous fields of application like tablet binder, drug delivering agent, coating material, bio-adhesive, etc. It can be expected that the current rate of development in this field will yield the next generation a remarkable number of highly available efficient materials for different pharmaceutical applications.

KEYWORDS

- okra gum
- pharmaceutical applications
- tablet excipients

REFERENCES

Adebayo, A. S., & Itiola, O. A., (2003). Effects of breadfruit and cocoyam starch mucilage binders on disintegration and dissolution behaviors of paracetamol tablet formulations. *Pharmaceutical Technology, 27*(3), 78–78.

Alamri, M. S., (2014). Okra-gum fortified bread: Formulation and quality. *J. Food Technol., 51*(10), 2370–2381.

Alamri, M. S., Mohamed, A. A., & Hussain, S., (2013). Effects of alkaline-soluble okra gum on rheological and thermal properties of systems with wheat or corn starch. *Food Hydrocolloid, 30*(2), 541–551.

Alamri, M. S., Mohamed, A., Hussain, S., & Xu, J., (2012). Effect of Okra extract on properties of wheat, corn and rice starches. *J. Food Agric Environ., 10*(1), 217–222.

Attama, A. A., Adikwu, M. U., & Amorha, C. J., (2003). Release of indomethacin from bioadhesive tablets containing carbopol 941 modified with Abelmoschus esculentus (okra) gum. *Bollettino Chimico Farmaceutico, 142*(7), 298–302.

Avachat, A. M., Dash, R. R., & Shrotriya, S. N., (2011). Recent investigations of plant-based natural gums, mucilages and resins in novel drug delivery systems. *Ind. J. Pharm. Edu. Res., 45*(1), 86–99.

Bakre, L. G., & Jaiyeoba, K. T., (2009a). Evaluation of a new tablet disintegrant from dried pods of Abelmoschus esculentus L (Okra). *Asian Journal of Pharmaceutical and Clinical Research, 2*(3), 83–91.

Bakre, L. G., & Jaiyeoba, K. T., (2009b). Effects of drying methods on the physicochemical and compressional characteristics of okra powder and the release properties of its metronidazole tablet formulation. *Arch. Pharm. Res., 32*(3), 259.

Baveja, S. K., Rao, K. V., & Arora, J., (1988). Examination of natural gums and mucilages as sustaining materials in tablet dosage forms. *Indian J. Pharm. Sci., 50*(2).

Bhat, U. R., & Tharanathan, R. N., (1987). Functional properties of okra (Hibiscus esculentus) mucilage. *Starch-Starke, 39*(5), 165–167.

Bhattacharyya, N. (2012). An antioxidant-rich fermented substrate produced by a newly isolated bacterium showing antimicrobial property against human pathogen, maybe a potent nutraceutical in the near future. *World Science Publisher, 1*(2).

Bhosale, R., Gangadharappa, H. V., Moin, A., Gowda, D. V., & Osmani, A. M., (2015). Grafting technique with special emphasis on natural gums: Applications and perspectives in drug delivery. *The Natural Products Journal, 5*(2), 124–139.

Chanchal, D. K., Alok, S., Kumar, M., Bijauliya, R. K., Rashi, S., & Gupta, S. (2018). A Brief Review on Abelmoschus Esculentus Linn. Okra. *International Journal of Pharmaceutical Sciences and Research, 9*(1), 58–66.

Costandino, A. J., & Romanchick-Cerpovicz, J. E., (2004). Okra polysaccharides as egg white substitute. *J. Am. Diet Assoc., 104*, 44–48.

Emeje, M. O., Isimi, C. Y., & Kunle, O. O., (2007). Evaluation of okra gum as a dry binder in Paracetamol tablet formulations. *Continental J. Pharm. Sci., 1*, 15–22.

Farinde, A. J., Owolarafe, O. K., & Ogungbemi, O. I., (2007). An overview of production, processing, marketing and utilization of okra in egbedore local government area of Osun State, Nigeria. *Agricultural Engineering International: CIGR Journal, IX.*

Ghori, M. U., Alba, K., Smith, A. M., Conway, B. R., & Kontogiorgos, V., (2014). Okra extracts in pharmaceutical and food applications. *Food Hydrocolloid, 42*, 342–347.

Gt, K., Gowthamarafan, K., Rao, B. G., & Suresh, B., (2002). Evaluation of binding properties of *Plantago ovata* and *Trigonella foenumgraecum iviucilages. Indian Drugs, 39*, 8.

Horisawa, E., Komura, A., Danjo, K., & Otsuka, A., (1993). Effect of binder characteristics on the strength of agglomerates prepared by the wet method. *Chem. Pharm. Bull., 41*(8), 1428–1433.

Hu, S. M., & Lai, H. S., (2017). Developing low-fat banana bread by using okra gum as a fat replacer. *Journal of Culinary Science & Technology, 15*(1), 36–42.

Ikoni, O., (2014). Some physicochemical properties of acetaminophen pediatric suspensions formulated with okra gums obtained from different extraction processes as a suspending agent. *Asian Journal of Pharmaceutics (AJP): Free Full-Text Articles from Asian J Pharm., 5*(1).

Jadhav, N., Gaikwad, V., Nair, K., & Kadam, H., (2009). Glass transition temperature: Basics and application in pharmaceutical sector. *Asian Journal of Pharmaceutics, 3*(2), 82.

Kalu, V. D., Odeniyi, M. A., & Jaiyeoba, K. T., (2007). Matrix properties of a new plant gum in controlled drug delivery. *Arch Pharm. Res., 30*(7), 884–889.

Kwok, T. S. H., Sunderland, B. V., & Heng, P. W. S., (2004). An investigation on the influence of a vinyl pyrrolidone/vinyl acetate copolymer on the moisture permeation, mechanical and adhesive properties of aqueous-based hydroxypropyl methylcellulose film coatings. *Chem. Pharm. Bull., 52*(7), 790–796.

Malviya, R., (2011). Extraction characterization and evaluation of selected mucilage as pharmaceutical excipient. *Polymers in Medicine, 41*(3), 39–44.

Momoh, M. A., Akikwu, M. U., Ogbona, J. I., & Nwachi, U. E., (2008). In vitro study of release of metronidazole tablets prepared from okra gum, gelatin gum and their admixture. *Bio-Research, 6*(1), 339–342.

Nasipuri, R. N., Igwilo, C. I., Brown, S. A., & Kunle, O. O., (1997). Mucilage from Abelmoschus esculentus fruits-a potential pharmaceutical raw material. Part 11-emulsyly properties. *J. Pharm. Res. Devt., 2*(1), 27–34.

Ndjouenkeu, R., Akingbala, J. O., & Oguntimein, G. B., (1997). Emulsifying properties of three African food hydrocolloids: Okra (Hibiscus esculentus), dika nut (Irvingia gabonensis), and khan (Belschmiedia sp.). *Plant Food Hum Nutri., 51*(3), 245–255.

Nep, E. I., & Conway, B. R., (2010). Characterization of grewia gum, a potential pharmaceutical excipient. *Journal of Excipients and Food Chemicals, 1*(1), 30–40.

Newton, A. M. J., Indana, V. L., & Kumar, J., (2015). Chronotherapeutic drug delivery of Tamarind gum, chitosan and okra gum controlled release colon targeted directly compressed propranolol HCl matrix tablets and in-vitro evaluation. *Int. Jo Biol. Macro.*, *79*, 290–299.

Ofoefule, S. I., & Chukwu, A., (2011). Application of *Abelmoschus esculentus* gum as a mini-matrix for furosemide and diclofenac sodium tablets. *Indian J Pharm. Sci.*, *63*(6), 532.

Ogaji, I., & Nnoli, O., (2010). Film coating potential of okra gum using paracetamol tablets as a model drug. *Asian Journal of Pharmaceutics*, *4*(2), 130.

Okoye, E. I., Onyekweli, A. O., & Kunle, O. O., (2011). Okra gum-an economic choice for the amelioration of capping and lamination in tablets. *Annals of Biological Research*, *2*(2), 30–42.

Onunkwo, G. C., (1996). Physical properties of sodium salicylate tablets formulated with Abelmoschus esculentus gum as binder. *Acta Pharmacologica.*, *46*, 101–107.

Pawan, P., & Nitin, K., (2013). Design and development of sustained release matrix tablet of diclofenac sodium using natural polymer. *International Research Journal of Pharmacy, 4*, 169–176.

Prajapati, V. D., Jani, G. K., Moradiya, N. G., & Randeria, N. P., (2013). Pharmaceutical applications of various natural gums, mucilages and their modified forms. *Carbohydrate Polymers, 92*(5), 1685–1699.

Qasem, A. A. A., Alamri, M. S., Mohamed, A. A., Hussain, S., Mahmood, K., & Ibraheem, M. A., (2017). Effect of okra gum on pasting and rheological properties of cake-batter. *J. Food Meas.*, *11*(2), 827–834.

Rhodes, C. T., & Porter, S. C., (1998). Coatings for controlled-release drug delivery systems. *Drug Dev. Ind. Pharm.*, *24*(12), 1139–1154.

Romanchik-Cerpovicz, J. E., Costantino, A. C., & Gunn, L. H., (2006). Sensory evaluation ratings and melting characteristics show that okra gum is an acceptable milk-fat ingredient substitute in chocolate frozen dairy dessert. *J. Acad. Nutr. Diet.*, *106*(4), 594–597.

Romanchik-Cerpovicz, J. E., Tilmon, R. W., & Baldree, K. A., (2002). Moisture retention and consumer acceptability of chocolate bar cookies prepared with okra gum as a fat ingredient substitute. *J. Acad. Nutr. Diet.*, *102*(9), 1301–1303.

Romanchik-Cerpovicz, J. E., Tilmon, R. W., & Baldree, K. A., (2002a). Moisture retention and consumer acceptability of chocolate bar cookies prepared with okra gum as a fat ingredient substitute. *J. Acad. Nutr. Diet.*, *102*(9), 1301–1303.

Romanchik-Cerpovicz, J. E., Tilmon, R. W., & Baldree, K. A., (2002b). Moisture retention and consumer acceptability of chocolate bar cookies prepared with okra gum as a fat ingredient substitute. *J. Am. Diet Assoc.*, *102*(9), 1301–1303.

Rowe, R. C., Sheskey, P. J., & Owen, S. N. C., (2002). *Handbook of Pharmaceutical Excipients* (6th edn., pp. 1–794). Pharmaceutical Press: London.

Roy, A., Khanra, K., Mishra, A., & Bhattacharyya, N., (2012). General analysis and antioxidant study of traditional fermented drink Handia, its concentrate and volatiles. *Advances in Life Science and Its Applications, 1*(3), 54–57.

Sharma, N., Kulkarni, G. T., & Sharma, A., (2013). Development of Abelmoschus esculentus (Okra)-based mucoadhesive gel for nasal delivery of rizatriptan benzoate. *Tropical Journal of Pharmaceutical Research, 12*(2), 149–153.

Sinha, P., Ubaidulla, U., & Nayak, A. K., (2015a). Okra (Hibiscus esculentus) gum-alginate blend mucoadhesive beads for controlled glibenclamide release. *Int. J. Biol. Macromol., 72*, 1069–1075.

Sinha, P., Ubaidulla, U., Hasnain, M. S., Nayak, A. K., & Rama, B., (2015b). Alginate-okra gum blend beads of diclofenac sodium from aqueous template using ZnSO4 as a cross-linker. *Int. J. Biol. Macromol., 79*, 555–563.

Symecko, C. W., & Rhodes, C. T., (1995). Binder functionality in tabletted systems. *Drug Dev. Ind. Pharm., 21*(9), 1091–1114.

Tavakoli, N., Ghassemi, D. N., Teimouri, R., & Hamishehkar, H., (2008). Characterization and evaluation of okra gum as a tablet binder. *Jundishapur Journal of Natural Pharmaceutical Products, 3*(1), 33–38.

Zaharuddin, N. D., Noordin, M. I., & Kadivar, A., (2014). The use of *Hibiscus esculentus* (Okra) gum in sustaining the release of propranolol hydrochloride in a solid oral dosage form. *Bio. Med. Research International, 2014.*

CHAPTER 9

Pharmaceutical Applications of Fenugreek Seed Gum

DILIPKUMAR PAL,[1] PHOOL CHANDRA,[2] NEETU SACHAN,[2]
MD SAQUIB HASNAIN,[3] and AMIT KUMAR NAYAK[4]

[1]*Department of Pharmaceutical Sciences, Guru Ghasidas Vishwavidyalaya (A Central University), Bilaspur – 495009, C.G, India*

[2]*School of Pharmaceutical Sciences, IFTM University, Lodhipur Rajput, Delhi Road (NH-24), Moradabad–244102, UP, India*

[3]*Department of Pharmacy, Shri Venkateshwara University, NH-24, Rajabpur, Gajraula, Amroha – 244236, U.P., India*

[4]*Department of Pharmaceutics, Seemanta Institute of Pharmaceutical Sciences, Mayurbhanj – 757086, Odisha, India*

9.1 INTRODUCTION

The term 'gum' refers to polysaccharide hydrocolloids, which do not form a part of cell wall, but are exudates or slims and are pathological products (Prajapati et al., 2013). Plant gums are defined as those substances of plant origin that are obtained as exudations from the fruits, trunks, or branches of trees spontaneously after the mechanical injury of the plant by the incision of the barks or by the removal of a branch or after invasion by bacteria or fungi (Hamden et al., 2010; Jones and Smith, 1949). The plant gums are amorphous substances containing carbon, hydrogen, and oxygen, and they are members of the carbohydrate group. In many cases, small amounts of nitrogen are detectable, and this may be traceable to the proteinaceous debris arising from the enzyme, which is responsible for the formation of the gums or it may arise from contact of the gum with protein material of the tree. The plant polysaccharides are hydrophilic substances and are characterized by dissolving in cold water or taking up water to form mucilage (Hamden et al., 2010).

9.2 FENUGREEK GUM

9.2.1 SOURCE

Fenugreek is known as *Trigonella foenum-graecum* belongs from the family *Leguminosae* is indigenous to the Mediterranean region, the northern part of Africa, the western part of Asia and Canada. The seeds of fenugreek have been employed in food and remedy as a component for many years (Brummer et al., 2003). Fenugreek seeds contain high percentages of carbohydrates, mainly polysaccharides. Fenugreek seed gum (FSG) is a soluble fiber obtained from the seeds of a legume plant, *Trigonella foenum-graecum,* which is commonly grown worldwide and predominantly in India. In a work, Brummer et al., (2003) extracted FSG as per the scheme depicted in Figure 9.1. Extraction procedure for fenugreek gum was found augmented in order to attain the low content of protein. Preliminary tests illustrated that during extraction, a lower temperature will yield the gums with less protein contaminants and significantly upsetting the yield. Therefore, a defatted, deactivated fenugreek seed produces FSG, which is optimized at 10°C for 2 h to provide a yield of 22% along with 2.36% protein contaminants (Table 9.1). Pronase was used to purify FSG, which further reduced protein contaminates to the level of 0.5% (Table 9.2) without upsetting the molecular weight of galactomannans. By HPSEC, changes in molecular weight were supervised.

9.2.2 CHEMISTRY

FSG, a complex polysaccharide, is made up of galactose and mannose (galactomannan). FSG consists of a $(1{\rightarrow}4)$-β-D-mannan backbone to which single α-D-galactopyranosyl groups are attached at the O-6 position of the D-mannopyranosyl residues (Brummer et al., 2003; Nayak et al., 2018). FSG has been shown to blunt rises in blood glucose and cholesterol postprandially as well as regulate the production of cholesterol in the liver (Srinivasan, 2006). FSG has also exhibited hypoglycemic effects, especially in persons and animals with type 1 and type 2 diabetes mellitus (Hannan et al., 2007).

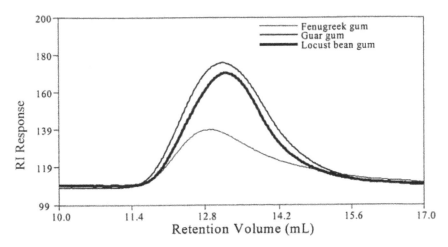

FIGURE 9.1 High-performance size exclusion chromatogram of fenugreek gum (fraction B), guar gum, and LBG. (Reprinted with permission from Brummer, Cui and Wang, 2003. © Elsevier.)

TABLE 9.1 Classification of Gums

S. No.	Basis	Class	Example
1.	Charge	Non-ionic gums	Guar gum, locust bean gum, tamarind gum, xanthan gum
		Anionic gums	Gum arabic, karaya gum, gellan gum, carrageenans
2.	Shape	Short branch	Xanthan gum, guar gum
		Branch on branch	Gum Arabic, tragacanth gum
3.	Origin	Seed gums	Guar gum, karaya gum, ipomoea, fenugreek, locust bean gum, premcem gum, lesquerellafendleri gum
		Plant exudates	Chicle gum, konjac, gum Arabic, gum ghatti, gum karaya, acacia gum, tragacanth
		Microbial exudates	Dextran, gellan gum, xanthan gum, tara gum, spruce gum
		See weed	Sodium alginate, alginic acid, carrageenans, agar–agar

Reprinted with permission from Prajapati, Jani, Moradiya, and Randeria (2013). © 2013 Elsevier.

TABLE 9.2 Composition of Fenugreek Seed and Fenugreek Seed Galactomannan

Fenugreek seed		
Lipid		7.24 ± 0.42
Protein		34.10 ± 0.84
Galactomannan		22.57 ± 2.8
Total ethanol soluble sugars		8.06 ± 0.62
Fenugreek seed galactomannan (%, w/w)		
Protein		
	Fraction A	2.36 ± 0.04
	Fraction B	0.57 ± 0.03
G/M ratio[a]		
	Fraction A	1.00/1.02
	Fraction B	1.00/1.05

All measurements are on a dry weight basis.

[a]Determined by HPLC–PAD analysis of acid hydrolyzed gum.

(Reprinted with permission from Brummer, Cui and Wang, 2003. © Elsevier.)

9.2.3 PROPERTIES

In a research by Ciucanu and Kerek (1984), FSG was subjected to the methylation analysis and according to the method reported earlier in the literature with some modification. The molecular weight of FSG was also determined along with other reported gums like locust bean gum (LBG) and guar gum with the help of HPSEC coupled with refractive index, viscosity, and right angle laser light-scattering detectors. Intrinsic viscosity of identical was estimated by using HPSEC data with a system of the triple detector. The chromatogram from the HPSEC analysis of FSG, guar gum, and LBG is shown in Figure 9.2. Comparison of average molecular weight of FSG, guar gum, and LBG are given in Table 9.3. Figure 9.3 shows the mechanical spectra of a 1% fenugreek gum solution. The surface tension concentration dependence for FSG fraction A and B is presented in Figure 9.4. All these galactomannans exhibited the capability to lessen the interfacial tension at the interface of tetradecane–water (Figures 9.5 and 9.6), and the magnitude of this reduction being concentration dependent. The linkage patterns of FSG fraction B are shown in Table 9.4. The steady shear viscosity of FSG was obtained (2, 1, 0.5, 0.2 and 0.1% solutions) along with the solutions of guar gum and LBG in a concentration of 1 and 0.5%. The storage and loss modulus (i.e., G' and G"), as well as complex viscosity (viscoelastic properties), were observed within the linear viscoelastic region (2% strain) for 0.5, 1, and 2% FSG solutions and 1% solutions for LBG and guar gum. The surface and interfacial tension were also recorded using Surface Tensionmat 21 at 21–22°C employing the Du Nouy ring method (Brummer et al., 2003).

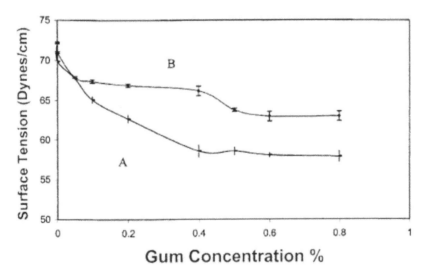

FIGURE 9.2 Changes of surface tension of fenugreek gum fraction A and B with gum concentration at 21–22°C. . (Reprinted with permission from Brummer, Cui and Wang, 2003. © Elsevier.)

TABLE 9.3 Molecular Weight, Intrinsic Viscosity and Radius of Gyration for Fenugreek, Guar, and LBG

Gum	Mw	Intrinsic viscosity (dl/g)	Rg (nm)
LBG	1,198,667	14.38	82.88
Guar	1,303,607	10.5	76.64
Fenugreek	1,418,000	9.61	75.08

(Reprinted with permission from Brummer, Cui and Wang, 2003. © Elsevier.)

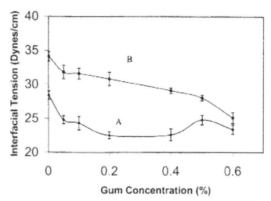

FIGURE 9.3 Concentration dependence of interfacial tension for fenugreek gum (fraction A and B) at tetradecane/water interface (21–22°C) (Reprinted with permission from Brummer, Cui and Wang, 2003. © Elsevier.).

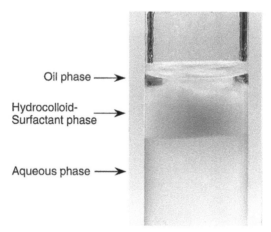

FIGURE 9.4 Photomicrographs of the 'three phases' separate from O/W emulsions stabilized with fenugreek gum, after centrifugation of 1500 ^{x}g. The 'in between phase' – the emulsifier rich phase can be detected between the aqueous and the oil phases. (Reprinted with permission from (Garti, Madar, Aserin and Sternheim, 1997. © Elsevier.).

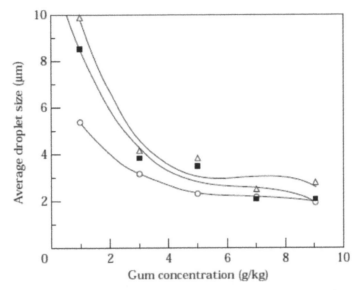

FIGURE 9.5 Average droplet size of 50 g/kg *n*-tetradecane-in-water emulsion stabilized with fenugreek gum as emulsifier, at different aging times. (○) = 0 h; (■) = 1 h; (Δ) = 168 h (Reprinted with permission from Garti, Madar, Aserin and Sternheim, 1997. © Elsevier.).

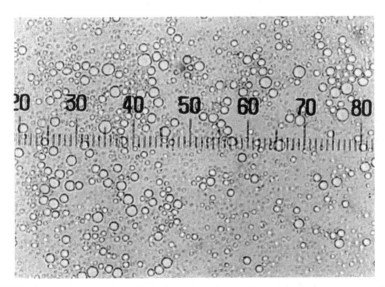

FIGURE 9.6 Typical droplets of 50 g/kg *n*-tetradecane-in-water emulsions prepared with 9 g/kg fenugreek gum after homogenization with ultra-turrax and microfluidizer (1 scale unit = 2.5 μm). (Reprinted with permission from (Garti, Madar, Aserin and Sternheim, 1997. © Elsevier.)

TABLE 9.4 Chemical Name and Deduced Linkage of PMAA of Fenugreek Gum

Chemical name	Relative abundance (%)	Deduced linkage
1,5-di-O-acetyl-(1-deuterio)-2,3,4,6-tetra-O-methyl hexitol	42	t-Gal$p$
1,4,5-tri-O-acetyl-(1-deuterio)-2,3,6-tri-O-methyl hexitol	12.7	4-Man$p$
1,4,5,6-tetra-O-acetyl-(1-deuterio)-2,3-di-O-methyl hexitol	45.2	4,6-Man$p$
1,5,6-tri-O-acetyl-(1-deuterio)-2,3,4-tri-O-methyl hexitol	T[a]	t-6-Man$p$

[a] Trace.

Reprinted with permission from Brummer, Cui and Wang, 2003. © Elsevier.

9.3 APPLICATIONS

9.3.1 EFFECT ON DIABETES

FSG (galactomannan) has already been reported for its inhibitory effect on the digestive enzymes related to diabetes, hyperlipidemia, and liver-kidney dysfunctions (Hamden et al., 2010). Hamden et al., (2010) performed the experiments on diabetic rats (glycemia $\geq$ 2 g/L). They induced diabetes intraperitoneally by administering a nicotinamide solution (1,000 mg/kg) that was followed 90 min later by streptozotocin solution (150 mg/kg). They administered FSG (galactomannan) for two months to diabetic rats. After two-month treatment, they found that the supplementation of FSG for surviving diabetic rats was found to significantly decrease the activities of intestine maltase, lactase, and sucrase by 39, 47, and 48%, respectively. Again, the same rats reduced glucose concentration in plasma by 56% as compared to the diabetic untreated rats. FSG (galactomannan) also revert the activity of lipase in intestine back by 46% as compared to diabetic rats along with a decrease in cholesterol and triglycerides in plasma. This gum also reverted the increased level of Liver dysfunction indices (AST, ALT, and LDH) by 38, 109, and 129%, respectively along with in decrease of kidney dysfunction indices (urea and creatinine rates in plasma) by 56 and 50%, respectively (Hamden et al., 2010).

9.3.2 PREPARATION OF SOY PROTEIN ISOLATE (SPI) AND FSG DISPERSED SYSTEMS

Food hydrocolloids show an important role in adjusting the viscosity and elasticity (rheological properties) of liquid and solid food products. Soy proteins are used as functional ingredients in food manufacturing due to increased demand by virtue of their role in human nutrition and health (Utsumi et al., 1997). Polysaccharides are cast off in admixture to proteins mainly to improve the stability of dispersed systems (Dickinson, 2003). Gelling properties and other functional properties of proteins are being modified in the presence of hydrocolloid gums (Hua et al., 2003). The gelation behavior of protein-polysaccharide mixtures generally falls into three patterns, i.e., the formation of covalent bonds between two polymers; polyanion-polycation electrostatic interactions; and formation of composite gel due to the mutual exclusion of each component (Morris, 1990).

Scientist (Hefnawy and Ramadan, 2011) prepared the six portions of Soy protein isolate (SPI) (1 g) and 0.3, 0.5, 1, 2, 3 or 4 g of FSG) were mixed in a dry state manually to prepare the respective SPI-FSG blends in the weight ratios of SPI/FSG = 3:1, 2:1, 1:1, 1:2, 1:3 and 1:4. Six of SPI-FSG dispersions were prepared by transferring the blends into 0.1 M sodium phosphate buffer (pH 7.4) while stirring with a magnetic stirrer for 60 min at ambient temperature to provide dispersion with 0.1% soy protein. These dispersions were used for determination of viscosity and emulsifying properties. Based on preliminary data, SPI-FSG dispersions in the weight ratio of 1:2, which gave the optimum emulsifying properties were chosen to evaluate the effects of pH, heat, and salt treatments. Nitrogen solubility appears to be the most important factor affecting the emulsifying activity (EA) of proteins (Halling, 1981). Generally, protein solubility has positively correlated with EA. SPI-FSG emulsions were equally stable over a wide range of NaCl concentrations (0.1 to 1 M, pH 7.4). For SPI-FSG, the excellent compatibility between these two biopolymers over a wide range of NaCl concentrations indicated that the affinity between the biopolymers was non-electrostatic. The excellent stability of SPI-FG emulsions over the range of pH 3.0 to pH 9.0 indicates that FSG probably formed a protective layer around SPI coated droplets. This resulted in enhancing the stabilizing properties of SPI based emulsion and prevented SPI perspiration at isoelectric point (IP). The EA and emulsifying stability (ES) of SPI-FSG were stable after heat treatment at 85°C for 1 h.

9.3.3 AS FOOD EMULSIFIER

Bipurified FSG (1–9 g/kg) was utilized to emulsify *n*-Tetradecane (50 g/kg) (Garti et al., 1997). They were homogenized either by a homogenizer with variable speed of 8000–24,000 rpm or a microfluidizer. Droplet size distribution was measured using a Coulter counter. Evaluation of the amount of gum adsorbed (Cads) onto the oil droplets was performed by centrifugation of the O/W emulsion at medium speed (1000 × g) until full separation into three phases of oil, water, and the emulsifier phase was obtained. Considerably smaller droplets were observed as compared to those originated from guar gum. Plot between the average sizes of droplet as the function of FSG concentration and the time for storage demonstrate that at low concentrations of gum and partial coverage of droplet, the droplets are large and have a propensity to unite rapidly with time whereas an augmentation in the concentration of gum at the equivalent level of oil permits improved droplets coverage resulting in the development of smaller droplets with merely a trivial inclination to combine.

9.3.4 GELATINIZATION AND RETROGRADATION BEHAVIORS STARCH/FSG COMPOSITE SYSTEM

A team of scientists studied the gelatinization and retrogradation behaviors of corn starch in an aqueous system at a reasonably lower concentration of starch (5%) in the presence of FSG (Funami et al., 2008a). During gelatinization, FSG addition (0.5% w/v) augmented the peak viscosity of the composite system (5% starch) when coil overlap parameter was C[Z] >2.38 and Mw > 87.0 × 10^4 g/mol (average molecular weight) of the gum, shifting the onset of viscosity increase at lower temperatures.

Another group again studied gelatinization and retrogradation behaviors of corn starch in an aqueous system in the presence of FSG with a range of molecular weights (7.5 ×10^4 to 20.7 ×10^5 g/mol) and having Rg (z-average root-mean-square radius of gyration) in between 16.3–122.3 nm. A composite system containing 15% starch showed peak viscosity during gelatinization when C[Z] (C: gum concentration) and the molecular weight of the gum were larger than 2.38 and 87.0×10^4 g/mol, correspondingly, were increased on the addition of FSG (0.5 w/v %). Whereas, it reduced the rate constant of the composite system which confirmed the association in relation to rheological creep conformity and storage time at 4°C for 14 days (Funami et al., 2008b).

9.3.5 FOR IMPROVING EMULSION STABILITY

To advance the emulsification properties, scientists prepared conjugates of soy whey protein isolate (SWPI)-FSG (hydrolyzed as well as unhydrolyzed) by using Maillard-type reaction in a controlled dry state circumstance (60°C, 75% relative humidity for 3 days). To study the effect of molecular weight on emulsifying properties, this partly hydrolyzed FSG using 0.05 M HCl for 10 min (HD10), 30 min (HD30) and 50 min (HD50) at a temperature of 90°C. With the help of sodium dodecyl sulfate-polyacrylamide gel electrophoresis (SDS-PAGE), SWPI-FSG conjugates formation was confirmed. Again, analysis of average particle size and particle size distribution propose that conjugation of SWPI-FSG at 60°C for 3 days was adequate to yield comparatively smaller size droplets in the oil-in-water type of emulsions (o/w) (Figure 9.7). In 1:3 and 1:5 ratio, SWPI: FSG was more efficient in stabilizing emulsion in comparison to 1:1 ratio. Unhydrolyzed FSG conjugates revealed improved emulsifying properties in comparison to partly hydrolyzed FSG conjugates. Lowering of the particle size of emulsions followed the given order: SWPI-unhydrolyzed FSG > SWPI-HD10 > SWPI-HD30 > SWPI-HD50 was (Kasran et al., 2013a). Due to the presence of both hydrophobic regions and charged hydrophilic regions, many proteins work as highly effective emulsifiers for lowering the surface tension as well as interact with the interface of the emulsion. Nevertheless, at the IP of the protein, this property of emulsification might be vanished (Kasran et al., 2013b). Retention of molecular integrity and solubility is the main advantage of the covalent complexes of protein-polysaccharide conjugate over non-covalent complexes for a varied range of experimental (Dickinson and Galazka, 1991). These conjugate possess stability over changes in temperature, ionic strength, and pH along with functional properties viz. enhanced emulsifying properties (Dickinson and Euston, 1991; Schmitt et al., 1998). These conjugates also possess improved functional properties, including enhanced emulsifying properties (Akhtar and Dickinson, 2007; Shepherd et al., 2000), improved solubility predominantly at pH around the IP of the protein (Chevalier et al., 2001; O'Regan and Mulvihill, 2009) and augmented heat stability against to protein itself (Aoki et al., 1999; Chevalier et al., 2001).

A team of scientists (Kasran et al., 2013b) studied the aptness of SWPI-FSG conjugates formulated by means of controlled dry heating of SWPI-FSG mixtures in stabilized o/w emulsion model systems and conjugation influence on the solubility of protein as a function of pH. They investigated the protein-polysaccharide complexes in emulsions in an acidic medium having high salt concentration as well as applying heat on the conjugates prior to emulsification. The protein solubility for SWPI, SWPI-FSG mixture, and conjugates of SWPI-FSG as a function of pH (3.0 to 8.0) was shown in Figure 9.8. With Coomassie brilliant blue, SDS-PAGE of SWPI-FSG mixture, as well as conjugates, were stained for proteins, which showed poly-dispersed bands at the top of separating gel in conjugates signifying the development of products of high molecular weight (Figure 9.9). A small average droplet size was produced for emulsion prepared with SWPI-FSG conjugates. The results indicate that the conjugate is a highly effective stabilizer of o/w emulsion under this condition. In comparison, preparation of emulsion with SWPI-FSG conjugate at pH 7.0 also produced small initial emulsion droplets (1.01 ± 0.03 µm) (Kasran et al., 2013b). Emulsions prepared with heated SWPI-unhydrolyzed FSG conjugates solution at 75°C and 85°C showed a significant decrease of average particle sizes compared to unheated conjugates with an average diameter of about 1.4 µm. Figure 9.10 compares average droplet sizes of emulsions stabilized by SWPI and SWPI-FSG conjugates in the presence or absence of 0.5 M NaCl.

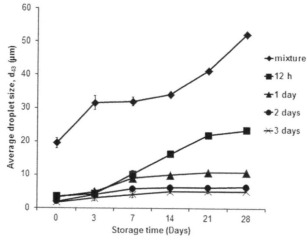

FIGURE 9.7   Comparison of average droplet size (d_{43}) for emulsions of canola oil (10 vol % oil, 0.8% emulsifier) stabilized by SWPI: HD30 (1:3) mixture and conjugates incubated at 60°C for 12 h, 1 day, 2 days and 3 days as a function of storage time from 0 to 28 days at pH 4.0 and stored at 25°C. (Reprinted with permission from Kasran, Cui and Goff, 2013a. © 2013 Elsevier.)

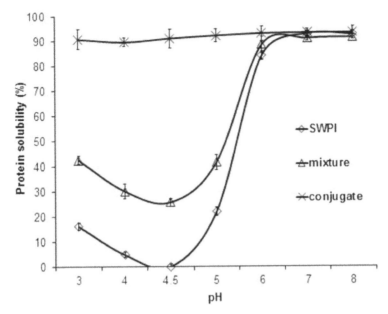

FIGURE 9.8 Protein solubility of SWPI, SWPI-fenugreek gum mixture, and SWPI-fenugreek conjugates as a function of pH from 3.0 to 8.0 at 22°C. (Reprinted with permission from Kasran, Cui and Goff, 2013b. © 2013 Elsevier.)

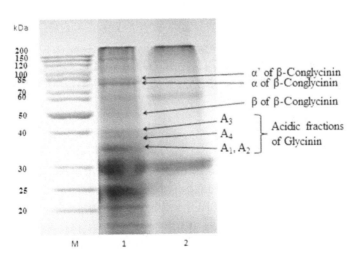

FIGURE 9.9 SDS-PAGE electrophoretogram of SWPI, SWPI-fenugreek gum mixture, and SWPI-fenugreek gum conjugate. The labeled lanes are (M) protein molecular weight standards; (1) SWPI-fenugreek gum mixture; and (2) SWPI-fenugreek gum conjugate. (Reprinted with permission from Kasran, Cui and Goff, 2013b. © 2013 Elsevier.)

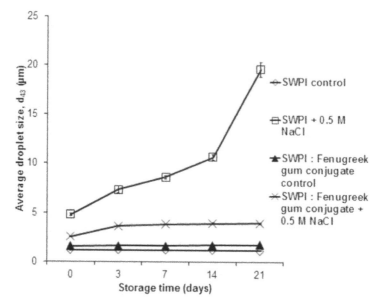

FIGURE 9.10 Comparison of average droplet size (d_{43}) for emulsions of canola oil (10 vol % oil, 0.8% emulsifier, pH 7.0) stabilized by SWPI (◊); SWPI ± 0.5 M NaCl (□); SWPI: fenugreek gum (1:3) conjugate (60°C, 3 days) (▲) and SWPI: fenugreek gum (1:3) conjugate (60°C, 3 days) ± 0.5 M NaCl (×) as a function of storage time from 0 to 21 days and stored at 25°C. (Reprinted with permission from Kasran, Cui and Goff, 2013b. © 2013 Elsevier.)

9.3.6 CARBOXYMETHYL FSG (GALACTOMANNAN)-GELLAN GUM-CALCIUM SILICATE COMPOSITE BEADS FOR GLIMEPIRIDE DELIVERY

Glimepiride, a sulphonylurea, belongs to the BCS class II shows poor aqueous solubility and limited dissolution rate at gastrointestinal (GI) lumen produces an incredible difference in oral bioavailability and poor clinical response with a short half-life (Ahmed et al., 2016). A mesoporous inorganic substance [$2CaO.3SiO_2.mSiO_2.nH_2O$], called as calcium silicate, has been unambiguously recognized for its assorted drug delivery uses. Calcium silicate brings the properties of tunable pore structure, excellent adsorption, and extremely high surface area. The hydrophobic drug molecules, typically fastened by electrostatic interfaces and hydrogen bonding links within the porous structure of calcium silicate, would show enhanced solubility and stability with controlled drug release profile (Han et al., 2013). The surface of calcium silicate can be tailored with FSG, a biopolymers in order to augment biocompatibility and targetability. To improve water solubility, drug carrier characteristics, and bioactivities of FSG, Carboxymethylation could be a promising method (Wang et al., 2016). Unfortunately, the high solubility and swelling features of carboxymethyl FSG in the aqueous fluid could typically result in poor encapsulation and premature elution of the entrapped drug (Bera et al., 2015b). The blending of two natural polymers could offer the most valuable approaches to improve the mechanical strength and performance. Gellan gum carries negative charges due to abundant glucuronic acid units and could be ionically crosslinked with bivalent and trivalent cations (e.g., aluminum, zinc, calcium, etc.) to yield rigid gellan gum hydrogel matrices (Nayak and Pal, 2014; Prezotti et al., 2014). Ca^{+2}-crosslinked FSG-blended gellan gum beads are reported in the literature (Nayak and Pal, 2014). Keeping these all thing, the group of scientist (Bera et al., 2018), developed gellan gum blended carboxymethyl FSG-calcium silicate based organic-inorganic composites cross-linked with an array of cations (i.e., $Ca^{+2}/Zn^{+2}/Al^{+3}$) and explored the drug carrier properties. They formulated mucoadhesive glimepiride loaded composite beads by ionotropic gelation method utilizing $Ca^{+2}/Zn^{+2}/Al^{+3}$ ions as cross-linkers. The properties of polymer-blend (gellan gum: carboxymethyl FSG) ratios and crosslinker types and calcium silicate inclusion on drug entrapment efficiency and cumulative drug release at 8 h were documented. Also, the morphological characteristics, polymer interaction-drug, and *in vivo* antidiabetic activity of these novel carriers were carried out. Synthesis of the carboxymethyl FSG was carried out by the Williamson synthesis *via* two-step process (I and II) (Singh et al., 2014).

$$ROH + NaOH \rightarrow RONa + H2O \qquad (I)$$

$$RONa + ClCH2COONa \rightarrow RO-CH2COONa + NaCl \qquad (II)$$

Estimation of the degree of substitution of carboxymethyl FSG was made with the previously described method (Gong et al., 2012) and was found to be 0.71 on the basis of FT-IR, DSC, and P-XRD studies. The SEM photograph of the composite beads encapsulating glimepiride (F-6) portrayed an approximately spherical shaped structure with distinctive large wrinkles, cracks, etc. scattered throughout the surface (Figure 9.11). The matrices also displayed a tight surface with greater irregularity, which could be due to the interfacial interactions between the calcium silicate molecules and biopolymer chains (Puttipipatkhachorn et al., 2005). The drug encapsulation efficiency, drug release, drug release data treatment, swelling index, water penetration velocity (Bera et al., 2015a), and *ex vivo* mucoadhesion evaluations were carried out. They resulted that among various matrices, the formulation reinforced with calcium silicate (F-6) depicted maximum drug encapsulation ability (96.74 ± 0.59%) and maximal drug release at 8 h (93.91 ± 1.06%) with mucoadhesivity. The optimized composite beads (F-6) was evaluated by Bera et al., (2018) for the effectiveness for diabetes, and that was induced in rats (200–250 g) by administrating streptozotocin (40 mg/kg body weight; i.p.) dissolved in freshly prepared 0.01 M citrate buffer (pH 4.5). Initial drug-induced hypoglycemic mortality due to streptozotocin in animals was prearranged by 20% glucose solution for a period of 24 h. Rats with marked hyperglycemia (fasting blood glucose range of 300–350 mg/dl) after 96 h of streptozotocin treatment, were used for the study. They were randomly divided into three groups (*n* = 6). Group I named diabetic control and administered 1 ml distilled water, group II administered suspension of glimepiride (2.5 mg/kg) and group III administered suspension of composite beads (F-6) containing the equivalent amount of glimepiride. All treatment was given orally to rats. Retro-orbital venous plexus was used for the blood collection at different time intervals for analysis of blood glucose employing commercial glucose kit by the oxidase-peroxidase method. Blood glucose level as percentage reduction versus time was recorded (Figure 9.12).

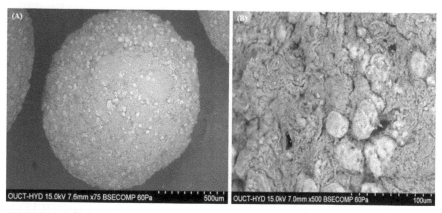

FIGURE 9.11   Scanning electron microphotograph of GLI-loaded nanocomposite beads (F-6) with less zoom (75 X) (A) and high zoom (500 X) (B). (Reprinted with permission from Bera, Mothe, Maiti, and Vanga, 2018. © 2017 Elsevier.)

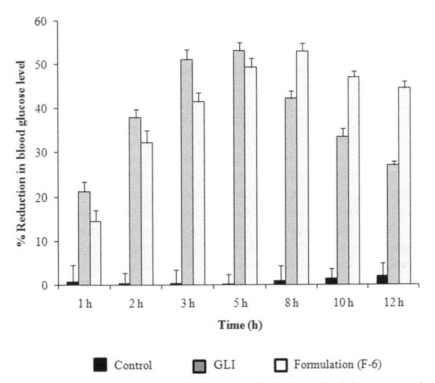

FIGURE 9.12 Comparative *in vivo* % reduction in blood glucose levels in streptozotocin-induced diabetic rats of group I (control), group II receiving pure GLI and group III receiving nanocomposite beads (F-6) containing GLI. Results are presented as mean ± SD; SD denoted by error bars. (Reprinted with permission from Bera, Mothe, Maiti, and Vanga, 2018. © 2017 Elsevier.)

9.3.7 FOR PREPARATION OF FSG BASED ACTIVE NANOCOMPOSITE FILM REINFORCED WITH NANOCLAYS

Plastics, basically a petroleum product shows serious drawbacks due to their non-degradability (Bordes et al., 2009). To avoid them and to fulfill their alternative, a biodegradable and environment-friendly material need to be searched. Drawbacks in biodegradable polymers can be modified by the application of nanotechnology and suggested to improve their characteristics taking into consideration the cost of them (Azeredo, 2009; Rhim et al., 2013; Sorrentino et al., 2007). In layered silicates, the surface of interlayer galleries is covered with interchangeable Ca^{+2} and Na^{+2}, allowing them non-compatibility and hydrophilic behavior with hydrophobic polymers. Organophilization is carried out to enable biocompatibility with polymers by increasing the galleries and exchanging the surface cations with various cations (Memiş et al., 2017). Researchers prepared FSG/clay nanocomposite films by solution casting method with Nano clays (Na$^+$ montmorillonite [MMT], halloysite [HNT] and Nanomer®I.44 P [NM]) at different amounts (0, 2.5, 5.0 and 7.5 g clay/100 g FG) and characterized (Memiş et al., 2017). The oxygen barrier and thermal properties are significantly ($P < 0.05$) reduced by an increasing amount of nanoclay. Agar diffusion tests guided that FSG based nanocomposite films showed strong antimicrobial properties to foodborne pathogens namely *Listeria monocytogenes, Escherichia coli, Staphylococcus aureus,* and *Bacillus cereus.* In the case of mechanical properties, nanoclay incorporation up to 5% provided higher ($P < 0.05$) tensile strength properties while elongation at break values of the films significantly ($P < 0.05$) decreased in the presence of clay in the film matrix. SEM micrographs showed that especially lower levels (up to 5%) of nanoclay reinforcements provided a homogeneous and smooth film structure. In conclusion, Overall, the findings of the study were promising that FG based nanocomposite films could be taken into consideration in food supplements packaging applications.

9.3.8   FOR SYNTHESIS OF FSG CARBAMATE

FSG was subjected to react with urea at high temperature in the solid state to yield carbamate derivatives and purified via dissolution in water, filtration followed by extraction in Soxhlet using 75% ethyl alcohol (Ragheb et al., 2015). Solubility, nitrogen content, and rheological properties were evaluated for the purified products. The nitrogen content increases from 1.45 to 4.13 by increasing the amount of urea from 20–100 g/100 g of the dry gum. The solubility of the carbamate derivatives in water was found to be depending on the extent of reaction expressed as % Nitrogen and acquire 1.45% Nitrogen or less water soluble, while those contain higher % Nitrogen are water-insoluble but soluble in 1% NaOH solution. All of the samples are characterized by non-Newtonian pseudoplastic behavior regardless of the % Nitrogen. The stability or the solvent used and apparent viscosity at any specific rate of shear was found to be dependent on nitrogen content.

9.4 CONCLUSION

As we are living in the modern era and utilizing the machines, electronic gadgets, and taking medicines for each and every failure of the physiological process. The chemically synthesized medicine produces so many side effects and to get rid of these problems, we should explore the raw materials as well as plant-derived products. Also, one can think for the search of widely abundant, inexpensive, nontoxic, naturally renewable, and biodegradable plant-derived materials. Among various natural plant products, fenugreek gum possesses so many characteristics, and that can be modified with the help of other substance for better drug delivery. Natural gums are used in pharmaceuticals for their diverse properties and applications. Among so many gums, FSG has so many properties and application for pharmaceutical use. This can be utilized for the development of gelatinization and retrogradation of a composite system with different concentration of corn. It can also be utilized as a natural emulsifier with less toxic level. FSG can also be utilized as food emulsifier for the preparation of carboxymethyl FSG (galactomannan)-gellan gum-calcium silicate composite beads for glimepiride delivery, active nanocomposite film reinforced with nanoclays and for the synthesis of FSG carbamate. This chapter will be helpful in applying appropriate strategies or achieving the desired properties of the composite system with the help of FSG.

9.5 FUTURE PROSPECTS

The vast applications of FSG suggest that these biomaterials have an optimistic and commercial future in various industrial fields. Therefore, it is judicious to expect that fenugreek gum continues to transform the advancements of green tools in the near future.

9.6 SUMMARY

Natural gums are used in pharmaceuticals for their diverse properties and applications. Among so many gums, FSG has so many properties and application for pharmaceutical use. This can be utilized for the development of gelatinization and retrogradation of a composite system with different concentration of corn. It can also be utilized as a natural emulsifier with less toxic level. FSG can also be utilized as food emulsifier for the preparation of composite beads, active nanocomposite film reinforced with nanoclays and for the synthesis of FSG carbamate. This chapter will be helpful in applying appropriate strategies or achieving the desired properties of the composite system with the help of FSG.

KEYWORDS

- **fenugreek gum**
- **galactomannan**
- **natural gum**
- ***Trigonella foenum-graecum***

REFERENCES

Ahmed, O. A. A., Zidan, A. S., & Khayat, M., (2016). Mechanistic analysis of Zein nanoparticles/PLGA triblock in situ forming implants for glimepiride. *International Journal of Nanomedicine, 11*, 543–555.

Akhtar, M., & Dickinson, E., (2007). Whey protein–maltodextrin conjugates as emulsifying agents: An alternative to gum Arabic. *Food Hydrocolloids, 21*(4), 607–616.

Aoki, T., Hiidome, Y., Kitahata, K., Sugimoto, Y., Ibrahim, H. R., & Kato, Y., (1999). Improvement of heat stability and emulsifying activity of ovalbumin by conjugation with glucuronic acid through the Maillard reaction. *Food Research International, 32*(2), 129–133.

Azeredo, H. M. C. D., (2009). Nanocomposites for food packaging applications. *Food Research International, 42*(9), 1240–1253.

Bera, H., Boddupalli, S., & Nayak, A. K., (2015a). Mucoadhesive-floating zinc-pectinate–sterculia gum interpenetrating polymer network beads encapsulating ziprasidone HCl. *Carbohydrate Polymers, 131*, 108–118.

Bera, H., Boddupalli, S., Nandikonda, S., Kumar, S., & Nayak, A. K., (2015b). Alginate gel-coated oil-entrapped alginate–tamarind gum–magnesium stearate buoyant beads of risperidone. *International Journal of Biological Macromolecules, 78*, 102–111.

Bera, H., Mothe, S., Maiti, S., & Vanga, S., (2018). Carboxymethyl fenugreek galactomannan-gellan gum-calcium silicate composite beads for glimepiride delivery. *International Journal of Biological Macromolecules, 107*, 604–614.

Bordes, P., Pollet, E., & Avérous, L., (2009). Nano-biocomposites: Biodegradable polyester/nanoclay systems. *Progress in Polymer Science, 34*(2), 125–155.

Brummer, Y., Cui, W., & Wang, Q., (2003). Extraction, purification and physicochemical characterization of fenugreek gum. *Food Hydrocolloids, 17*(3), 229–236.

Carpita, N. C., & Shea, E. M., (1989). Linkage structure of carbohydrates by gas chromatography-mass spectrometry (GC–MS) of partially methylated alditol acetates. In: Bierman, C. J., & McGinnis, G. D., (eds.), *Analysis of Carbohydrates by GLC and MS* (pp. 157–216). Boca Raton: CRC Press.

Chevalier, F., Chobert, J. M., Popineau, Y., Nicolas, M. G., & Haertlé, T., (2001). Improvement of functional properties of β-lactoglobulin glycated through the Maillard reaction is related to the nature of the sugar. *International Dairy Journal, 11*(3), 145–152.

Ciucanu, I., & Kerek, F., (1984). A simple and rapid method for the permethylation of carbohydrates. *Carbohydrate Research, 131*(2), 209–217.

Dickinson, E., & Euston, S. R., (1991). Stability of food emulsions containing both protein and polysaccharide. In: Dickinson, E., (ed.), *Food Polymers, Gels and Colloids* (pp. 132–146), Woodhead Publishing.

Dickinson, E., & Galazka, V. B., (1991). Emulsion stabilization by ionic and covalent complexes of β-lactoglobulin with polysaccharides. *Food Hydrocolloids, 5*(3), 281–296.

Dickinson, E., (2003). Hydrocolloids at interfaces and the influence on the properties of dispersed systems. *Food Hydrocolloids, 17*(1), 25–39.

Funami, T., Kataoka, Y., Noda, S., Hiroe, M., Ishihara, S., Asai, I., Takahashi, R., Inouchi, N., & Nishinari, K., (2008a). Functions of fenugreek gum with various molecular weights on the gelatinization and retrogradation behaviors of corn starch—2: Characterizations of starch and investigations of corn starch/fenugreek gum composite system at a relatively low starch concentration, 5w/v %. *Food Hydrocolloids, 22*(5), 777–787.

Funami, T., Kataoka, Y., Noda, S., Hiroe, M., Ishihara, S., Asai, I., Takahashi, R., & Nishinari, K., (2008b). Functions of fenugreek gum with various molecular weights on the gelatinization and retrogradation behaviors of corn starch—1: Characterizations of fenugreek gum and investigations of corn starch/fenugreek gum composite system at a relatively high starch concentration, 15w/v %. *Food Hydrocolloids, 22*(5), 763–776.

Garti, N., Madar, Z., Aserin, A., & Sternheim, B., (1997). Fenugreek Galactomannans as Food Emulsifiers. *LWT – Food Science and Technology, 30*(3), 305–311.

Gong, H., Liu, M., Chen, J., Han, F., Gao, C., & Zhang, B., (2012). Synthesis and characterization of carboxymethyl guar gum and rheological properties of its solutions. *Carbohydrate Polymers, 88*(3), 1015–1022.

Hamden, K., Jaouadi, B., Carreau, S., Bejar, S., & Elfeki, A., (2010). Inhibitory effect of fenugreek galactomannan on digestive enzymes related to diabetes, hyperlipidemia, and liver-kidney dysfunctions. *Biotechnology and Bioprocess Engineering, 15*(3), 407–413.

Han, Y., Zeng, Q., Li, H., & Chang, J., (2013). The calcium silicate/alginate composite: Preparation and evaluation of its behavior as bioactive injectable hydrogels. *Acta Biomaterialia, 9*(11), 9107–9117.

Hannan, J. M. A., Ali, L., Rokeya, B., Khaleque, J., Akhter, M., Flatt, P. R., & Abdel-Wahab, Y. H. A., (2007). Soluble dietary fiber fraction of Trigonella foenum-graecum (fenugreek) seed improves glucose homeostasis in animal models of type 1 and type 2 diabetes by delaying carbohydrate digestion and absorption, and enhancing insulin action. *British Journal of Nutrition, 97*(3), 514–521.

Hefnawy, H. T. M., & Ramadan, M. F., (2011). Physicochemical characteristics of soy protein isolate and fenugreek gum dispersed systems. *Journal of Food Science and Technology, 48*(3), 371–377.

Hua, Y., Cui, S. W., & Wang, Q., (2003). Gelling property of soy protein–gum mixtures. *Food Hydrocolloids, 17*(6), 889–894.

Jones, J. K. N., & Smith, F., (1949). Plant gums and mucilages. In: Pigm, W. W., & Wolfro, M. L., (eds.), *Advances in Carbohydrate Chemistry* (Vol. 4, pp. 243–291), Academic Press.

Kasran, M., Cui, S. W., & Goff, H. D., (2013a). Covalent attachment of fenugreek gum to soy whey protein isolate through natural Maillard reaction for improved emulsion stability. *Food Hydrocolloids, 30*(2), 552–558.

Kasran, M., Cui, S. W., & Goff, H. D., (2013b). Emulsifying properties of soy whey protein isolate–fenugreek gum conjugates in oil-in-water emulsion model system. *Food Hydrocolloids, 30*(2), 691–697.

Memiş, S., Tornuk, F., Bozkurt, F., & Durak, M. Z., (2017). Production and characterization of a new biodegradable fenugreek seed gum based active nanocomposite film reinforced with nanoclays. *International Journal of Biological Macromolecules, 103*, 669–675.

Morris, E. R., (1990). Mixed polymer gels. In: Harris, P., (ed.), *Food Gels*, (pp. 291–359). Dordrecht: Springer Netherlands.

Nayak, A. K., & Pal, D., (2014). *Trigonella foenum-graecum* L. seed mucilage-gellan mucoadhesive beads for controlled release of metformin HCl. *Carbohydrate Polymers, 107*, 31–40.

Nayak, A. K., Bera, H., Hasnain, M. S., & Pal, D., (2018). Synthesis and characterization of graft copolymers of plant polysaccharides. In: Thakur, V. K., (ed.), *Biopolymer Grafting*, (pp. 1–62), Elsevier.

O'Regan, J., & Mulvihill, D. M., (2009). Preparation, characterization and selected functional properties of sodium caseinate–maltodextrin conjugates. *Food Chemistry, 115*(4), 1257–1267.

Prajapati, V. D., Jani, G. K., Moradiya, N. G., & Randeria, N. P., (2013). Pharmaceutical applications of various natural gums, mucilages and their modified forms. *Carbohydrate Polymers, 92*(2), 1685–1699.

Prezotti, F. G., Cury, B. S. F., & Evangelista, R. C., (2014). Mucoadhesive beads of gellan gum/pectin intended to controlled delivery of drugs. *Carbohydrate Polymers, 113*, 286–295.

Puttipipatkhachorn, S., Pongjanyakul, T., & Priprem, A., (2005). Molecular interaction in alginate beads reinforced with sodium starch glycolate or magnesium aluminum silicate, and their physical characteristics. *International Journal of Pharmaceutics, 293*(1), 51–62.

Ragheb, A. A., EL-Rahman, A. A. A., Ibrahim, M. A., EL-Thalouth, I. A., & Al-Moaty, A. R. A., (2015). Synthesis and characterization of fenugreek gum carbamate. *Egyptian Journal of Chemistry, 58*(4), 403–414.

Rhim, J. W., Park, H. M., & Ha, C. S., (2013). Bio-nanocomposites for food packaging applications. *Progress in Polymer Science, 38*(10), 1629–1652.

Schmitt, C., Sanchez, C., Desobry-Banon, S., & Hardy, J., (1998). Structure and techno-functional properties of protein-polysaccharide complexes: A review. *Critical Reviews in Food Science and Nutrition, 38*(8), 689–753.

Shepherd, R., Robertson, A., & Ofman, D., (2000). Dairy glycoconjugate emulsifiers: Casein–maltodextrins. *Food Hydrocolloids, 14*(4), 281–286.

Singh, R., Maity, S., & Sa, B., (2014). Effect of ionic crosslink on the release of metronidazole from partially carboxymethylated guar gum tablet. *Carbohydrate Polymers, 106*, 414–421.

Sorrentino, A., Gorrasi, G., & Vittoria, V., (2007). Potential perspectives of bio-nanocomposites for food packaging applications. *Trends in Food Science & Technology, 18*(2), 84–95.

Srinivasan, K., (2006). Fenugreek (Trigonella foenum-graecum): A review of health beneficial physiological effects. *Food Reviews International, 22*(2), 203–224.

Utsumi, S., Matsumura, Y., & Mori, T., (1997). Structure-function relationships of soy protein. In: Damodaran, S., & Paraf, A., (eds.), *Food Proteins and Their Application,* (pp. 257–291). New York: Marcel Dekker.

Wang, Z. J., Xie, J. H., Shen, M. Y., Tang, W., Wang, H., Nie, S. P., & Xie, M. Y., (2016). Carboxymethylation of polysaccharide from *Cyclocarya paliurus* and their characterization and antioxidant properties evaluation. *Carbohydrate Polymers, 136,* 988–994.

Index